COLUMNS FOR GAS CHROMATOGRAPHY

BICENTENNIAL
1807
WILEY
2007
BICENTENNIAL

THE WILEY BICENTENNIAL—KNOWLEDGE FOR GENERATIONS

*E*ach generation has its unique needs and aspirations. When Charles Wiley first opened his small printing shop in lower Manhattan in 1807, it was a generation of boundless potential searching for an identity. And we were there, helping to define a new American literary tradition. Over half a century later, in the midst of the Second Industrial Revolution, it was a generation focused on building the future. Once again, we were there, supplying the critical scientific, technical, and engineering knowledge that helped frame the world. Throughout the 20th Century, and into the new millennium, nations began to reach out beyond their own borders and a new international community was born. Wiley was there, expanding its operations around the world to enable a global exchange of ideas, opinions, and know-how.

For 200 years, Wiley has been an integral part of each generation's journey, enabling the flow of information and understanding necessary to meet their needs and fulfill their aspirations. Today, bold new technologies are changing the way we live and learn. Wiley will be there, providing you the must-have knowledge you need to imagine new worlds, new possibilities, and new opportunities.

Generations come and go, but you can always count on Wiley to provide you the knowledge you need, when and where you need it!

WILLIAM J. PESCE
PRESIDENT AND CHIEF EXECUTIVE OFFICER

PETER BOOTH WILEY
CHAIRMAN OF THE BOARD

COLUMNS FOR GAS CHROMATOGRAPHY

Performance and Selection

Eugene F. Barry, Ph.D
Professor of Chemistry
University of Massachusetts Lowell

Robert L. Grob, Ph.D
Professor Emeritus, Analytical Chemistry
Villanova University

WILEY-
INTERSCIENCE

A JOHN WILEY & SONS, INC., PUBLICATION

Published by John Wiley & Sons, Inc., Hoboken, New Jersey.
Published simultaneously in Canada.

For general information on our other products and services or for technical support, please contact our Customer Care Department within the United States at (800) 762-2974, outside the United States at (317) 572-3993 or fax (317) 572-4002.

Wiley also publishes its books in a variety of electronic formats. Some content that appears in print may not be available in electronic formats. For more information about Wiley products, visit our web site at www.wiley.com.

Library of Congress Cataloging-in-Publication Data:

Barry, Eugene F.
  Columns for gas chromatography : performance and selection / Eugene F. Barry, Ph.D., Robert L. Grob, Ph.D.
      p. cm.
  Includes index.
  ISBN 978-0-471-74043-8
  1. Gas chromatography. I. Grob, Robert Lee. II. Title.
  QD79.C45B37 2007
  543'.85—dc22

                                                                    2006026963

Printed in the United States of America.

10 9 8 7 6 5 4 3 2 1

To our wives and families for their understanding and support during the many days of seclusion and confusion that we spent when completing this book.

Also, to our many students over the years, whom we hope have benefited from our dedication to the field of separation science.

It is my sad task to inform the reader that my good friend, colleague, and co-author, Dr. Robert L. Grob, passed away on October 22, 2006, several months after the manuscript associated with this was submitted to John Wiley & Sons. Dr. Grob made significant contributions to the field of chromatography and remains one of its most outstanding contributors and a very respected proponent. He was an excellent teacher, mentoring many students and encouraged many others to pursue chromatography and the discipline of analytical chemistry in general. He tirelessly gave much of his time to organizations such as the Eastern Analytical Symposium, Pittcon, and the Chromatography Forum of Delaware Valley. He is deeply missed, as are his welcoming smile and characteristic humorous laugh.

EUGENE F. BARRY

Nashua, New Hampshire
November 10, 2006

CONTENTS

4   Column Oven Temperature Control   **210**

PREFACE

The gas chromatographic column can be considered the heart of a gas chromatograph. As such, selection of a gas chromatographic column is made with the intended applications in mind and the availability of the appropriate inlet and detector systems. Over the past three decades the nature and design of columns have changed considerably from columns containing either a solid adsorbent or a liquid deposited on an inert solid support packed into a length of tubing to one containing an immobilized or cross-linked stationary phase bound to the inner surface of a much longer length of fused-silica tubing. With respect to packing materials, solid adsorbents such as silica gel and alumina have been replaced by porous polymeric adsorbents, while the vast array of stationary liquid phases in the 1960s have been reduced greatly in number, to a smaller number of phases of greater thermal stability. These became the precursors of the chemically bonded or cross-linked phases of today. Column tubing fabricated from copper, aluminum, glass, and stainless steel served the early analytical needs of gas chromatographers. In this book the performance of packed gas chromatographic columns is discussed for several reasons. To the best knowledge of the authors, no other text is available that treats packed column gas chromatography (GC). At the same time, there is a substantial subset of gas chromatographers who use packed columns, and the once-popular book by Walter Supina, *The Packed Column in Gas Chromatography,* has not been updated. Presently, fused-silica capillary columns 10 to 60 m in length in with an inner diameter of 0.20 to 0.53 mm are in widespread use. Furthermore, we believe additional strengths of the book are the extensive tabulation of USP methods in Chapter 2 and the handy list of column dimensions for ASTM, EPA, and NIOSH methods in Chapter 3. Appendix A consists of 160 packed column separations that once formed the red booklet *Packed Column Separations,* now Supelco's Brochure 890B. Our goal in including these separations on packed columns is to facilitate transfer of a packed column separation over to an appropriate capillary column with the aid of a column cross-reference chart or table.

Although GC may be viewed, in general, as a mature analytical technique, improvements in column technology, injection, and detector design appear steadily nonetheless. Innovations and advances in GC have been made in the last decade, with the merits of the fused-silica column as the focal point and have been driven primarily by the environmental, petrochemical, forensic, and toxicological fields as well as by advances in sample preparations and mass spectrometry. The cost of a gas chromatograph can range from $6000 to over $100,000, depending on the types and number of detectors, injection systems, and peripheral devices such as data systems, headspace and thermal desorption devices, pyrolyzers, and autosamplers. When one also factors in the regular purchase of high-purity gases required for operation of the chromatograph, it quickly becomes apparent that a sizable investment is required. For example, cost-effectiveness and good chromatographic practice dictate that users of capillary columns give careful consideration to column selection; otherwise, the entire gas chromatographic process may be compromised. This book provides the necessary guidance for column selection regarding dimensions of column length, inside and outside diameter, film thickness, and type of capillary column chosen with the injection system and detectors in mind. Properly implemented connections of the column to the injector and detector and the presence of high boilers, particulate matter in samples, and so on, are included for the interested reader.

Chromatographers have seen the results of splendid efforts by capillary column manufacturers to produce columns having lower residual activity and capable of withstanding higher column temperature operation with reduced column bleeding. With the increasing popularity of high-speed or fast GC and the increasing presence of GC-MS in the analytical laboratory, especially for environmental, food, flavor, and toxicological analyses, improvements in column performance that affect the MS detector have steadily evolved (i.e., columns with reduced column bleed). There is also an increased availability of capillary columns exhibiting stationary-phase tuned selectively for specific applications obtained by synthesis of new phases, blending of stationary phases, and preparation of phases with guidance from computer modeling. These advances and the chemistries associated with them are also surveyed. Additional special features found in this book are the advantages of computer assistance in gas chromatography, multidimensional GC, useful hints for successful GC, and GC resources on the Internet.

A comprehensive state-of-the-art treatment of column selection, performance, and technology such as this book should aid the novice with this analytical technique and enhance the abilities of those experienced in the use of GC.

Nashua, New Hampshire 2006   EUGENE F. BARRY

Malvern, Pennsylvania 2006   ROBERT L. GROB

ACKNOWLEDGMENTS

The authors are deeply grateful to Heather Bergman, Associate Science Editor at John Wiley, for her astute guidance and assistance as well as her gentle nudging during the completion of this book.

Many scientists have contributed to the book. The authors wish to acknowledge these scientists: Drs. Lindauer and O'Brien for the excellent job of compiling the information on USP methods in Table 2.15. Dr. Richard Lindauer has three decades of analytical R&D experience in pharmaceutical quality control. He consults in pharmaceutical and dietary supplement analyses, method development, validation, reference standards, USP–NF issues, regulatory issues, and laboratory operations. For 18 years he led analytical research at the *U.S. Pharmacopeia* as director of the R&D and drug research and testing labs. Dr. Matthew O'Brien was in pharmaceutical research and development with Merck Research Laboratories for twenty-five years and currently is a consultant on regulatory requirements, collectively known as chemistry manufacturing and controls, and is a consultant in quality systems with the Quantic Group, Ltd. As a consultant, Dr. O'Brien has participated on quality teams for major pharmaceutical companies and supported the filing of NDAs and INDs.

We are appreciative of the efforts of Dr. Rick Parmely at Restek for supplying us with Tables 3.13 and 3.14 and the gift of ProezGC software, and the assistance of Dr. Russel Gant and Ms. Jill Thomas of Supelco in arranging for us to include Brochure 890B in its entirety: the 160 packed column separations appearing in Appendix A. This booklet was once standard issue to those using packed columns. We also thank Dr. Dan Difeo and Anthony Audino of SGE for their assistance with tables and photographs. We are grateful to Pat Spink of ChemSW for the gift of the GC–SOS optimization package and to LC Resources (Drs. Lloyd Snyder, John Dolan, and Tom Jupille) for a gift of DryLab software. We are appreciative of the donation of the GC Racer from Dr. Steve MacDonald.

We wish to take this opportunity to thank the following persons for their assistance with this book as well as for providing instructional material for our short courses at Pittcon and EAS: Sky Countryman at Phenomenex; Joseph Konschnik, Christine Varga, and Mark Lawrence at Restek; Mark Robillard at Supelco; and Reggie Bartram at Alltech Associates.

Diane Goodrich deserves special thanks for her typing and word-processing skills in reformatting tables, as does G. Duane Grob for his professional computer assistance skills.

We are grateful to all of you.

1 Introduction

1.1 EVOLUTION OF GAS CHROMATOGRAPHIC COLUMNS

The gas chromatographs and columns used today in gas chromatography have evolved gradually over five decades, similar to the evolution and advancements made in the cars we drive, the cameras we use, and the television sets that we view. In retrospect, the first gas chromatographs may be considered rather large compared to the modern versions of today, but these were manufactured for packed columns. Also, the prevailing thinking of the day was that "bigger was better," in that multiple packed columns could be installed in a large column oven. This is not necessarily true in all cases today, as now we know that a large column compartment oven offers potential problems (e.g., thermal gradients, hot and cold spots) if a fused-silica capillary column is installed in a spacious oven. The columns used in the infancy of gas chromatography were prepared with metal tubing such as copper, aluminum, and stainless steel. Only stainless steel packed columns remain in use; columns fabricated from the more reactive metals copper and aluminum are no longer used, and the use of copper tubing in gas chromatography has basically been limited to carrier gas and detector gas lines and ancillary connections.

Packing of such columns proved to be an event, often involving two or more people and a stairwell, depending on the length to be packed. After uncoiling the metal tubing to the desired length and inserting a wad of glass wool into one end and attaching a funnel to the other end, packing material would be added gradually while another person climbed the stairs taping or vibrating the tubing to further settle the packing in the column. When no further packing could be added, the funnel was detached, a wad of glass wool inserted at that end, and the column coiled manually to the desired diameter. These tapping and vibration processes produced fines of packing materials and ultimately contributed to the overall inefficiency of the chromatographic process. Glass columns were soon recognized to provide an attractive alternative to metal columns, as glass offers a more inert surface texture, although these columns are more fragile, requiring careful handling; have to be configured in geometrical dimensions for the instrument in which they are to be installed; and the presence of silanol groups on the inner glass surface has to be addressed through silylation chemistries. Additional features of glass columns are that one can visualize how well a column is packed, the presence of any void regions, and the possible discoloration of the packing at the inlet end of the column due to the

Columns for Gas Chromatography: Performance and Selection, by Eugene F. Barry and Robert L. Grob
Copyright © 2007 John Wiley & Sons, Inc.

accumulation of high boilers and particulates, which indicates that it is time for fresh packing. Most laboratories today leave the preparation of packed columns to column vendors. A generation of typical packed columns fabricated from these materials are shown in Figure 1.1; packed columns are discussed in Chapter 2.

The evolution of the open-tubular or capillary column may be viewed as paralleling that of the packed column. The first capillary columns that demonstrated efficiency superior to that of their packed column counterparts were made primarily of stainless steel. Glass capillary columns gradually replaced stainless steel capillary columns and proved to offer more inertness and efficiency as well as less surface activity, but their fragility was a problem, requiring straightening of column ends followed by the addition of small aliquots of fresh coating solution. Perhaps the most significant advance in column technology occurred in 1979 with the introduction of fused silica by Hewlett-Packard (now Agilent Technologies) (1,2). Today, the fused-silica capillary column is in wide use and its features, such as superior inertness and flexibility, have contributed to concurrent improvements in inlet and detector modifications that have evolved with advances in stationary-phase technology. Because of the high impact of fused silica as a column material, resulting in excellent chromatography, numerous publications have focused on many aspects of this type of column. For example, the interested reader is referred to an informative review by Hinshaw, who describes how fused-silica capillary columns are made (3), and some guidance offered by Parmely, who has outlined how successful gas chromatography with fused-silica columns can be attained (4). A generation of capillary columns are shown in Figure 1.2; capillary columns are the subject of Chapter 3.

The first group of stationary phases were adsorbents, somewhat limited in number, for gas–solid chromatography with packed columns, and included silica gel, alumina, inorganic salts, molecular sieves, and later, porous polymers and graphitized carbons, to name a few. Today, porous-layer open tubular or PLOT columns employ these adsorbents as stationary phases where adsorbent particles adhere to the inner wall of fused-silica capillary tubing. However, more numerous were the number of liquids studied as liquid phases for gas–liquid chromatography. In 1975, Tolnai and co-workers indicated that more than 1000 liquids had been introduced as stationary liquid phases for packed columns up to that time (5); to state that almost every chemical in an organic stockroom has been used as a stationary liquid phase is probably not much of an exaggeration.

Some popular liquid phases in the early 1960s are listed in Table 1.1. The majority of these are no longer in routine use (exceptions being SE-30, Carbowax 20M, squalane, and several others) and have been replaced with more thermally stable liquids or gums. Also of interest in this list is the presence of Tide, a laundry detergent, and diisodecyl, dinonyl, and dioctyl phthalates; the phthalates can be chromatographed easily on a present-day column. From 1960 through the mid-1970s, a plethora of liquid phases were in use for packed column gas chromatography to provide the selectivity needed to compensate for the low efficiency of the packed column to yield a given degree of resolution. When classification schemes of liquid phases were introduced by McReynolds and Rohrschneider (see Chapter 2), the number of liquid phases for packed columns decreased gradually over time.

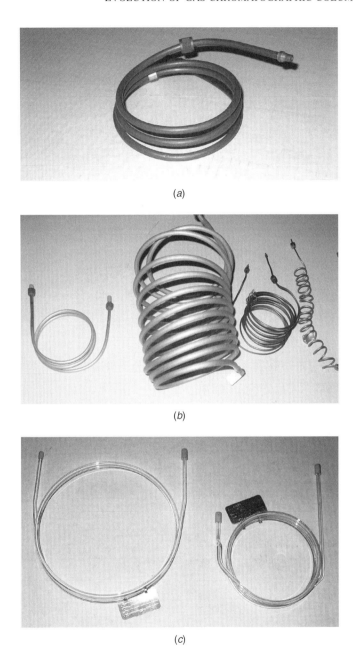

Figure 1.1   Various columns and materials used for packed column gas chromatography: (a) 6 ft × 0.25 in. o.d. copper tubing; (b) from left to right: 4 ft × 0.25 in. o.d. aluminum column, 20 ft × $\frac{3}{8}$ in. o.d. aluminum column for preparative GC, 10 ft × 1/8 in. o.d. stainless steel column, 3 ft × $\frac{1}{8}$ in. o.d. stainless steel column coiled in a "pigtail" configuration; (c) glass packed gas chromatographic columns, 2 m × 0.25 in. o.d. × 4 mm i.d. Note the differences in the length configuration of the ends, specific to two different chromatographs.

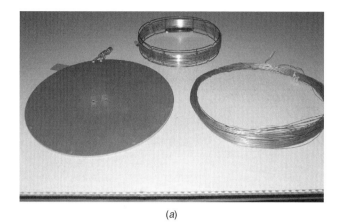

(a)

(b)

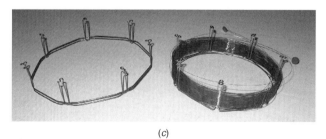

(c)

Figure 1.2   Various columns and materials employed for capillary gas chromatography: (a) left: 25 m × $\frac{1}{16}$ in. o.d. stainless steel capillary column in a "pancake" format, center: 30 m × 0.25 mm i.d. aluminum-clad fused-silica column, right: blank or uncoated stainless steel capillary tubing $\frac{1}{16}$ o.d.; (b) 60 m × 0.75 mm i.d. borosilicate glass capillary column for EPA method 502.2; (c) 30 m × 0.25 mm i.d. fused-silica capillary column; also pictured is a typical cage used to confine and mount a fused-silica column.

TABLE 1.1   Stationary Phases Used in Gas Chromatography Prior to 1962[a]

Liquid Phase	Maximum Temp. (°C)	Liquid Phase	Maximum Temp. (°C)
Inorganic eutectic mixtures	>350	Carbowax 1500	175–200
Silicone elastomer E301	300	*Diisodecyl phthalate*	175–180
Silicone rubber gum SE-30	250–350	*Dioctyl phthalate*	160
		Zinc stearate	160
DC high-vacuum grease	250–350	Nujol paraffin oil	150–200
Apiezon M	275–300	Sucrose acetate isobutyrate	150–200
Polyethylene	275–300		
Apiezon L	240–300	Apiezon N	150
Ethylene glycol–isophthalic acid polyester	280	Sorbitol-silicone oil X525	150
			150
		Bis(2-ethylhexyl) tetrachlorophthalate	150
Embaphase silicone oil	250–260		
Neopentyl glycol succinate	230–250	Polypropylene glycol	140–150
		7,8-Benzoquinoline	115–150
Carbowax 20M	220–250	*Dinonyl phthalate*	130–150
Polyphenyl ether	250	Carbowax 1000	125–190
Tide detergent	225–250	Bis(2-ethylhexyl) sebacate	125–175
Resoflex R446 and R447	240		
Polyester		Tricresyl phosphate	125–160
Diethylene glycol succinate (LAC-3-R728)	200–225	*Carbowax 600*	125–150
		Benzyldiphenyl	120–140
		Fluorene picrate	120
Cross-linked diethylene glycol adipate (LAC-2-R446)	200–225	Diglycerol	100–150
		THEED (tetrahydroxy-ethylethylenediamine)	100–135
		Carbowax 400	100–130
Ucon polyglycol LB-550-X	200–225	Squalane	80–140
Ucon 50 HB 2000	200–240	*Glycerol*	70–120
Carbowax 6000	200–225	*β, β′-Oxydipropionitrile*	50–100
Carbowax 4000	200–225	Dibenzyl ether	60–80
Carbowax 4000 monostearate	200–220	*Hexadecane*	40–60
Celanese ester No. 9	200	Tetraethylene glycol dimethyl ether	40–80
Convoil 20	200		
Nonylphenoxypoly(ethylene-oxy)ethanol	200	Propylene glycol–AgNO₃	40–50
		Di-*n*-butyl maleate	40–50
Convachlor-12	200	Dimethylsulfolane	35–40
Triton X-305	200	Quinoline–brucine	25
Reoplex 400	190–270	*Dimethylformamide*	0
DC silicone oil 550	190–225		
DC silicone oil 200	175–225		

Source: Data from ref. 6.
[a] Phases in italic type may be viewed as obsolete.

In this reduced number of phases, only a small fraction proved useful in capillary gas chromatography (GC), where thermal stability of thin films of stationary liquids at elevated temperatures and wettability of fused silica, for example, become key chromatographic issues. On the other hand, only a few stationary phases of relatively lower selectivity are needed in capillary GC because of the much higher efficiency of a capillary column. Industrial-grade lubricants such as the Apiezon greases and Ucon oils suitable for the packed column needs of the day were replaced by more refined synthetic or highly purified versions of polysiloxanes or polyethylene glycols. Polysiloxanes, for example, are one of the most studied classes of polymers and may be found as the active ingredients in caulks, window gaskets, contact lenses, and car waxes; the first footprints on the moon were made by polysiloxane footwear (7). Another well-studied class of polymers are the more polar polyethylene glycols (PEGs), which also have use in a variety of applications (e.g., one active component in solutions used in preparation for colonoscopy procedures is PEG 3550). However, as efficacious and effective as polysiloxanes and polyethylene glycols may be in these applications, many studies have shown that only those polysiloxanes and polyethylene glycols that have well-defined chemical and physical properties satisfy the requirements of a stationary phase for capillary GC, as discussed in Chapter 3.

The reader will find equations for the calculation of column efficiency, selectivity, resolution, and so on, in Chapter 2. Included among these equations is an expression for time of analysis, an important parameter for a laboratory that has a high sample throughput. Temperature programming of a column oven, operation of a gas chromatographic column at a high flow rate or linear velocity, selection of favorable column dimensions, and optimization of separations with computer assistance can all reduce analysis time. In the last decade, fast or high-speed GC has emerged as a powerful mode in gas chromatography and is treated in Chapter 4. As gas chromatography comes closer to becoming a mature analytical technique, one tends to focus on the present and may forget early meritorious pioneering efforts, particularly the role of temperature programming for fast gas chromatography. Such is the case with temperature programming in GC, introduced by Dal Nogare and his colleagues, the first proponents of its role in reducing the time of analysis (8,9). The first reported separation in fast GC and schematic diagrams of circuitry of the column oven are shown in Figure 1.3.

1.2 CENTRAL ROLE PLAYED BY THE COLUMN

The gas chromatographic column may be considered to be the central item in a gas chromatograph. Over the last three decades, the nature and design of the column has changed considerably from one containing either a solid adsorbent or a liquid deposited on an inert solid support packed into a length of tubing to one containing an immobilized or cross-linked stationary phase bound to the inner surface of a much longer length of fused-silica tubing. With respect to packing materials, as noted earlier, solid adsorbents such as silica gel and alumina have been replaced by porous polymeric adsorbents, and the vast array of stationary liquid phases in the 1960s was by the next decade reduced to a much smaller number of phases of greater thermal stability. These stationary phases became the precursors of the chemically bonded or cross-linked phases of today. Column tubing fabricated from copper, aluminum, glass, and stainless steel served the early analytical needs of gas chromatographers. Presently, fused-silica capillary columns 10 to 60 m in length and 0.20 to 0.53 mm in inner diameter are in widespread use.

Although gas chromatography may be viewed in general as a mature analytical technique, improvements in column technology, injection, and detector design appear steadily nonetheless. During the last decade, innovations and advancements in gas chromatography have been made with the merits of the fused-silica column as the focal point and have been driven primarily by the environmental, petrochemical,

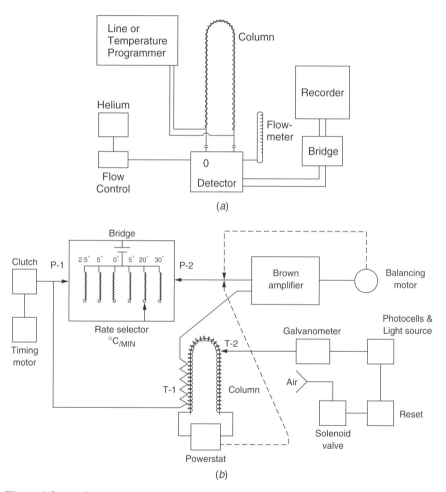

Figure 1.3 (*a*) Programmed-temperature apparatus constructed by Dal Nogare and Harden; (*b*) closed-loop proportional temperature controller; (*c*) programmed-temperature separation of a C5–C10 paraffin mixture: (A) heating rate 30°C/min, flow rate 100 mL/min, starting temperature 40°C, (B) same conditions except heating rate 5°C /min, (C) isothermal separation at 75°C, flow rate 100 mL/min. [From ref. 8. Reprinted with permission from *Anal. Chem.*, **31** 1829–1832 (1959). Copyright © 1959 American Chemical Society.] (*Continued*)

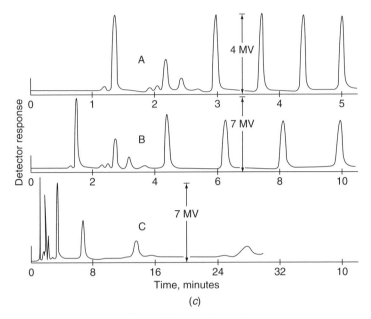

Figure 1.3 (*continued*)

and toxicological fields as well as by advances in sample preparations and mass spectrometry. Despite being a mature technology, there are three parallel tracks on which advancements in gas chromatography steadily appear. First, continued improvements in column technology have resulted in more efficient and thermally stable columns; second, advances in both electronics and pneumatics have provided impetus to fast GC, and third, sample preparation, mass spectrometry, and multidimensional GC represents three additional areas where great strides continue to be made. The interested reader may refer to the fourth edition of *Modern Practice of Gas Chromatography* for detailed coverage of all aspects of GC (10).

1.3 JUSTIFICATION FOR COLUMN SELECTION AND CARE

The cost of a gas chromatograph can range from $6000 to over $100,000, depending on the type and number of detectors, injection systems, and peripheral devices, such as a data system, headspace and thermal desorption units, pyrolyzers, and autosamplers. When one factors in purchase of the high-purity gases required for operation of the chromatograph, it quickly becomes apparent that a sizable investment has been made in capital equipment. For example, cost-effectiveness and good chromatographic practice dictate that users of capillary columns should give careful consideration to column selection. The dimensions and type of capillary column should be chosen with the injection system and detectors in mind, considerations that are virtually nonissues with packed columns. Careful attention should also be paid to properly implemented connections of the column to the injector and detector and the presence of high boilers, particulate matter in samples, and so on.

The price of a column ($200 to $800) may be viewed as relatively small compared to the initial, routine, and preventive maintenance costs of the instrument. In fact, a laboratory may find that the cost of a set of air and hydrogen gas cylinders of research-grade purity for FID (flame ionization detector) operation is far greater than the price of a single conventional capillary column! Consequently, to derive maximum performance from a gas chromatographic system, the column should be carefully selected for an application, handled with care following the suggestions of its manufacturer, and installed as recommended in the user's instrument manual.

The introduction of inert fused-silica capillary columns in 1979 markedly changed the practice of gas chromatography, enabling high-resolution separations to be performed in most laboratories (1,2). Previously, such separations were achieved with reactive stainless steel columns and with glass columns. After 1979, the use of packed columns began to decline. A further decrease in the use of packed columns occurred in 1983 with the arrival of the megabore capillary column of 0.53-mm inner diameter (i.d.), which serves as a direct replacement for a packed column. These developments, in conjunction with the emergence of immobilized or cross-linked stationary phases tailored specifically for fused-silica capillary columns and the overall improvements in column technology and affordability of mass spectrometry (MS), have been responsible for the wider acceptance of capillary GC.

Trends.   The results of a survey of 12 leading experts in gas chromatography appeared in 1989 and outlined their thoughts on projected trends in gas chromatographic column technology, including the future of packed columns versus capillary columns (11). Some responses of that panel are:

1. Packed columns are used for approximately 20% of gas chromatographic analyses.
2. Packed columns are employed primarily for preparative applications, for fixed gas analysis, for simple separations, and for separations for which high resolution is not required or not always desirable [e.g., polychlorinated biphenyls (PCBs)].
3. Packed columns will continue to be used for gas chromatographic methods that were validated on packed columns, where time and cost of revalidation on capillary columns would be prohibitive.
4. Capillary columns will not replace packed columns in the near future, although few applications require packed columns.

Shortly thereafter, in 1990, Majors summarized the results of a more detailed survey on column use in gas chromatography, this one, however, soliciting response from *LC/GC* readership (12). Some conclusions drawn from this survey include:

1. Nearly 80% of the respondents used capillary columns.
2. Capillary columns of 0.25- and 0.53-mm i.d. were the most popular, as were column lengths of 10 to 30 m.

3. The methyl silicones and polyethylene glycol stationary phases were preferred for capillary separations.

4. Packed columns were used primarily for gas–solid chromatographic separations such as gas analyses.

5. The majority of respondents indicated the need for stationary phases of higher thermal stability.

Majors conducted helpful GC user surveys again in 1995 (13) and 2003 (14). In the 2003 survey, the use of packed columns continued to decline because many packed column gas chromatographic methods have been replaced by equivalent capillary methods. There are now capillary column procedures for the U.S. Environmental Protection Agency (EPA), American Association of Official Analytical Chemists (AOAC), and *U.S. Pharmacopeia* (USP) methods. Despite the increase in capillary column users (91% in 2003 compared to 79% in 1990), there is still a significant number of packed column users, for several reasons: (1) packed columns and related supplies and accessories have a substantial presence in catalogs and Web sites of the major column vendors, and (2) the use of packed columns become apparent to the authors of this text after discussions with attendees in short courses on GC offered at professional meetings.

Other interesting findings in this 2003 survey included:

1. A pronounced increase in the use of columns of 0.10 to 0.18-mm i.d. Their smaller inner diameter permits faster analysis times and sensitivity, and their lower capacity is offset by the sensitive detectors available.

2. Columns of 0.2 to 0.25- and 0.32-mm i.d. in 20 to 30-m lengths continue to be the most popular.

3. 100% Methyl silicone, 5% phenylmethyl silicone, polyethylene glycol (WAX), and 50% phenylmethyl silicone continue to be the most popular stationary phases.

4. There appears to be a shift from gas–solid packed columns for the analysis of gases and volatiles to PLOT columns.

Column manufacturers rely on the current literature, the results of marketing surveys, the number of clicks on their Web sites, and so on, to keep abreast of the needs of practicing gas chromatographers. The fused-silica capillary column has clearly emerged as the column of choice for most gas chromatographic applications. A market research report covering 1993 (15) showed that $100 million was spent on capillary columns worldwide, and at an estimated average cost of $400 for a column, this figure represented about 250,000 columns. The number of columns and users has increased considerably since then, along with the cost of columns. Despite the maturity of capillary GC, instrument manufacturers continue to improve the performance of gas chromatographs, which has diversely extended the applications of gas chromatography.

Chromatographers can expect to see continued splendid efforts by capillary column manufacturers to produce columns that have lower residual activity and are capable of withstanding higher-column-temperature operation with reduced column bleeding. With the increasing popularity of high-speed or fast GC (Chapter 4) and the increasing presence of GC-MS in the analytical laboratory, especially for environmental, food, flavor, and toxicological analyses, improvements in column performance that affect the MS detector have steadily evolved, such as columns with reduced column bleed. There is also an increased availability of capillary columns exhibiting stationary-phase selectivity tuned for specific applications obtained by synthesis of new phases (16). For example, enhanced separation of the congeners of polychlorinated dibenzodioxin (PCDD), furan (PCDF), and PCBs can be achieved with the selectively tuned columns commercially available (17–21), discussed in more detail in Chapter 4. Interesting studies on blending stationary phases and phase preparation with guidance from computer modeling (22) and molecular simulation studies in gas–liquid chromatography have appeared in the literature (23,24).

1.4 LITERATURE ON GAS CHROMATOGRAPHIC COLUMNS

The primary journals in which developments in column technology and applications are published in hard-copy format and online versions include *Analytical Chemistry, Journal of Chromatography* (Part A), *Journal of Chromatographic Science, Journal of Separation Science* (formerly the *Journal of High Resolution Chromatography*, including the *Journal of Microcolumn Separations*), *and LC/GC* magazine. The biennial review issue of *Analytical Chemistry*, Fundamental Reviews (published in even-numbered years), contains concise summaries of developments in gas chromatography. An abundance of gas chromatographic applications may be found in the companion issue, Application Reviews (published in odd-numbered years), covering the areas of polymers, geological materials, petroleum and coal, coatings, pesticides, forensic science, clinical chemistry, environmental analysis, air pollution, and water analysis.

Most industrial and corporate laboratories as well as college and universities have access to literature searching through one of a number of online computerized database services (e.g., *SciFinder Scholar*). Although articles on gas chromatography in primary journals are relatively easy to locate, finding publications of interest in lesser known periodicals can be a challenge and is often tedious. *CA Selects, SciFinder Scholar*, and *Current Contents* are convenient alternatives. The biweekly *CA Selects*, Gas Chromatography topical edition available from Chemical Abstracts Service is a condensation of information reported throughout the world. *SciFinder Scholar* is a powerful searching capability, as it is connected to Chemical Abstracts Service but can retrieve information rapidly by either topic or author. *Current Contents*, in media storage format, provides weekly coverage of current research in the life sciences; clinical medicine; the physical, chemical, and earth sciences; and agricultural, biology, and environmental sciences.

With each passing year, the periodic commercial literature and annual catalogs of column manufacturers (in compact disk format from many column vendors) describing applications for their columns contain more and more useful technical information of a generic nature. In addition, we strongly recommend *LC/GC* as a valuable resource not only for timely technical articles but also "Column Watch" and a troubleshooting section for GC. However, the Internet has emerged as the most extensive source of chromatographic information in recent years, particularly the Web sites of column manufacturers.

1.5   GAS CHROMATOGRAPHIC RESOURCES ON THE INTERNET

The World Wide Web (WWW) has provided us with copious amounts of information through retrieval with search engines offered by an Internet service provider (ISP) (25). The Internet has affected our everyday activities with the convenience of communication by e-mail, online placement of orders for all types of items, and many other functions. There are numerous Web sites on gas chromatography in general, gas chromatographic columns, gas chromatographic detectors, and so on; all one has to do is locate them by "surfing the net." All manufacturers of gas chromatographic instrumentation, columns, and chromatographic accessories and supplies maintain Web sites and keep them updated. We strongly suggest that you identify and visit regularly the Internet addresses of column manufacturers, for example, and "bookmark" the corresponding Web sites. Internet addresses may change over time, as in the case of expansion or consolidation. For example, there has been some consolidation in the column industry for GC: J&W was purchased by Agilent Technologies, Chrompack by Varian, and Supelco by Sigma-Aldrich, and new column manufacturers (Phenomenex and VICI Gig Harbor Group) have entered the marketplace. It is thus impractical here to list the Web addresses of vendors, but "home pages" are easily searchable and updated continually, thus serving as an outstanding source of reference material for the practicing chromatographer. Lists of Web sites and addresses of vendors may be found in the annual *ACS Buyers' Guide* as well as in its counterparts in *LC/GC* and *American Laboratory*.

Listing all the gas chromatographic resources available on the Internet is not practical, but resourceful guides and information (often, in streaming video) that are available at gas chromatographic sites include:

- Information describing e-notes, e-newsletters, and e-seminars offered by a vendor
- Free downloads of software (e.g., retention time locking, method translation)
- Technical libraries of chromatograms searchable by solute or class of solute
- Column cross-reference charts
- Application notes
- Guides to column and stationary-phase selection
- Guides to selection of inlet liners
- Guides to column installation
- Guides to derivatization

- Troubleshooting guides
- Guides for syringe, septa, and ferrule selection
- Guides for setting up a gas chromatograph
- Past presentations at professional meetings such as Pittcon and EAS

Of the plethora of informative and significant ".com" and ".org." sites that are available, one deserves special mention because it serves as a path both for immediate assistance and for the continuing education of users of GC and HPLC: *the Chromatography Forum*, maintained by LC Resources (www.lcresources.com). There are several message boards (i.e., a GC message board, an LC message board, and several others) where one can post anonymously a chromatographic problem or question and others can post a response, initiating a dialogue on the topic. This site offers broadening of one's knowledge of the technique, even for the experienced user, and is a particularly valuable asset for an analyst working in an environment in which he or she is the sole chromatography user or does not have access to other resources or assistance with technical problems. Sometimes clickable links are overlooked as possibly noteworthy chromatographic sites; such sites may be embedded in a primary site selected by a search engine such as *Yahoo* or *Google*. An illustration of this is the site maintained by John Wiley, www.separationsnow.com, where there are several links, including one for Discussion Forums on GC, hyphenated techniques, HPLC, and others. Again, visit pertinent Web sites as part of the ongoing professional growth process of a chromatographer; also remember that a Web address may change sooner or later.

REFERENCES

1. R. Dandeneau and E. H. Zerenner, *J. High Resolut. Chromatogr. Chromatogr. Commun.*, **2**, 351 (1979).
2. R. Dandeneau and E. H. Zerenner, *Proc. 3rd International Symposium on Glass Capillary Gas Chromatography*, Hindelang, Germany, 1979, pp. 81–97.
3. J. V. Hinshaw, *LC/GC*, **9**, 1016 (2005).
4. R. Parmely, *LC/GC*, **11**, 1166 (2005).
5. G. Tolnai, G. Alexander, Z. Horvolgyi, Z. Juvancz, and A. Dallos, *Chromatographia*, **53**, 69 (2001).
6. S. Dal Nogare and R. Juvet, *Gas–Liquid Chromatography,* Wiley, New York, 1962, p. 121.
7. A. S. Abd-El-Aziz, C. E. Carraher, C. U. Pittman, J. E. Sheats, and M. Zeldin, *Macromolecules Containing Metal and Metal-like Elements*, Vol. 3, Wiley, Hoboken, NJ, 2004, p. xiii.
8. S. Dal Nogare and J. C. Harden, *Anal. Chem.*, **31**, 1829 (1959).
9. S. Dal Nogare and W. E. Langlois, *Anal. Chem.*, **32**, 767 (1960).
10. R. L. Grob and E. F. Barry (Eds.), *Modern Practice of Gas Chromatography,* 4th ed., Wiley, Hoboken, NJ, 2004.
11. R. E. Majors, *LC/GC*, **7**, 888 (1989).

12. R. E. Majors, *LC/GC*, **8**, 444 (1990).

13. R. E. Majors, *1995 Gas Chromatography User Study*, Advanstar Communications, Edison, NJ, 1995.

14. R. E. Majors, *2003 Gas Chromatography User Study,* Advanstar Communications, Iselin, NJ, 2003.

15. *Analy. Instrum. Ind. Rep.,* **10**(14), 4 (1993).

16. E. J. Guthrie and J. J. Harland, *LC/GC*, **12**, 80 (1994).

17. D. DiFeo, A. Hibberd, and G. Sharp, Pittsburgh Conference, Chicago, 2004, Poster 21900–600.

18. K. A. MacPherson, E. J. Reiner, T. K. Kolic, and F. L. Dorman, *Organohalogen Compound.*, **60**, 367 (2003).

19. T. Matsumura, Y. Masuzaki, Y. Seki, H. Ito, and M. Morita, *Organohalogen Compound.*, **60**, 375 (2003).

20. D. DiFeo, A. Hibberd, and G. Sharp, Pittsburgh Conference, Chicago, 2004, Poster 21900–500.

21. F. L. Dorman, G. B. Stidsen, C. M. English, L. Nolan, and J. Cochran, Pittsburgh Conference, Chicago, 2004, Poster 21900–200.

22. F. L. Dorman, P. D. Schettler, C. M. English, and D. V. Patwardhan, *Anal. Chem.*, **74,** 2133 (2002).

23. C. D. Wick, J. I. Siepmann, and M. R. Schure, *Anal. Chem.*, **74**, 37 (2002).

24. C. D. Wick, J. I. Siepmann, and M. R. Schure, *Anal. Chem.*, **74**, 3518 (2002).

25. G. I. Ouchi, *LC/GC*, **17**(4), 322 (1999).

2 Packed Column Gas Chromatography

2.1 INTRODUCTION

The first commercially available packed columns for gas chromatography were those available with the Perkin-Elmer vapor fractometer, Model 154, in 1954. Although the identities of the packings were at first proprietary, they soon became known to the scientific community. At first these columns were simply designated by a capital letter of the alphabet along with a brief description of the type(s) of analytes they could separate. Each column contained 20% liquid phase coated on 60/80-mesh Chromosorb. The columns and their chemical composition are given in Table 2.1.

Packed columns are still utilized for a variety of applications in gas chromatography. A packed column consists of four basic components: tubing in which packing material is placed; packing retainers (such as glass wool plugs or fritted metal plugs) inserted into the ends of the tubing to keep the packing in place; the packing material itself, which may be only a solid support or a solid support coated with a stationary phase (liquid substrate). The role and properties of solid support material, adsorbents, commonly used stationary phases and procedures for the preparation of packed columns are described in this chapter. Factors that can affect packed column performance are also discussed.

2.2 SOLID SUPPORTS AND ADSORBENTS

Supports for Gas–Liquid Chromatography

The purpose and role of the solid support is the accommodation of a uniform deposition of stationary phase on the surface of the support. The most commonly used support materials are primarily diatomite supports and graphitized carbon (which is also an adsorbent in gas–solid chromatography), and to a lesser extent, Teflon, inorganic salts, and glass beads. There is no perfect support material; each has limitations. Pertinent physical properties of a solid support for packed column GC are particle size, porosity, surface area, and packing density. Particle size affects column efficiency by means of an eddy diffusion contribution in the *van Deemter*

Columns for Gas Chromatography: Performance and Selection, by Eugene F. Barry and Robert L. Grob
Copyright © 2007 John Wiley & Sons, Inc.

TABLE 2.1   Perkin-Elmer Gas–Liquid Chromatographic Columns for the Model 154 Vapor Fractometer

Column	Liquid Substrate
A	Diisodecyl phthalate
B	Di-2-Ethylhexyl sebacate
C	Silicone oil (Dow Corning 200)
E	Dimethylsulfolane
N	Polyethylene glycol (Carbowax 1500)
O	Silicone grease (DC High-vacuum grease)
P	Polyethylene glycol succinate
Q	Apiezon
R	Polypropyene glycol (Ucon LB-550-X)

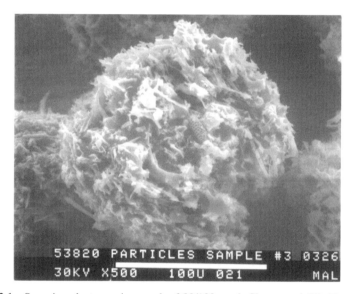

Figure 2.1   Scanning electron micrograph of 80/100-mesh Chromosorb W. (From ref. 1.)

expression:

$$H = 2\lambda d_p + \frac{2\gamma D_g}{u} + \frac{8}{\pi^2} \frac{k}{(1+k)^2} \frac{d_f^2}{D_l} u \tag{2.1}$$

Each term in Eq. 2.1 is discussed later in the chapter. The surface area of a support is governed by its porosity, the more-porous supports requiring greater amounts of stationary phase for coverage. A photomicrograph of 80/100-mesh Chromsorb W-HP appears in Figure 2.1, where the complex pore network is clearly evident.

Diatomite Supports.   Basically, two types of support are made from diatomite. One is pink and derived from firebrick, and the other is white and derived from filter aid. German diatomite firebrick is known as *Sterchmal*. Diatomite itself, diatomaceous

earth, is composed of diatom skeletons or single-celled algae that have accumulated in very large beds in numerous parts of the world. The skeletons consist
of a hydrated microamorphorous silica with some minor impurities (e.g., metallic oxides). The various species of diatoms number well over 10,000 from both
freshwater and saltwater sources. Many levels of pore structure in the diatom cell
wall cause these diatomites to have large surface areas (e.g., 20 m²/g). The basic
chemical differences between pink and white diatomites may be summarized as
follows:

1. White diatomite or filter aid is prepared by mixing it with a small amount of
   flux (e.g., sodium carbonate), followed by calcining (burning) at temperatures
   greater than 900°C. This process converts the original light-gray diatomite to
   white diatomite. The change in color is believed to be the result of converting
   the iron oxide to a colorless sodium iron silicate.
2. Pink or brick diatomite has been crushed, blended, and pressed into bricks,
   which are calcined at temperatures greater than 900°C. During the process
   the mineral impurities form complex oxides and silicates. The oxide of iron
   is the source of the pink color.

A support should have sufficient surface area so that the amount of stationary
phase chosen can be deposited uniformly and not leave active sites exposed on the
surface. Conversely, if excessive phase (above the upper coating limit of the support) is deposited on the support, the liquid phase may have a tendency to "puddle"
or pool on a support particle and can even spread to an adjacent particle, resulting
in a decrease in column efficiency due to unfavorable mass transfer of the analyte.
In Figure 2.2 a series of scanning electron micrographs of 20% Carbowax 20M
on 80/100-mesh Chromosorb W-HP are shown. A photomicrograph of a nonho-
mogeneous deposition of phase is shown in Figure 2.2a, where a large amount of
polymer distributed between two particles is visible in the left-hand portion of the
photograph. This packing ultimately yielded a column of low efficiency because of
unfavorable mass transfer, as opposed to the higher column efficiency associated
with a column packed with a more uniformly coated support (Figure 2.2b).

Pink firebrick supports, such as Chromosorb P and Gas Chrom R, are very strong
particulates that provide higher column plate numbers than those provided bar most
supports. Because of their high specific surface area, these supports can accommodate up to 30% percent loading of liquid phase, and their use is reserved for
the analysis of nonpolar species such as hydrocarbons. They must be deactivated,
however, when employed for the analysis of polar compounds such as alcohols
and amines. As a result, white filter-aid supports of lower surface area (e.g., Chromosorb W, Gas Chrom Q, and Supelcoport) are preferable, although they are more
fragile and permit a slightly lower maximum loading of about 25 wt% by weight
of liquid phase. A harder and improved support, Chromosorb G, is denser than
Chromosorb W but also exhibits a lower surface area and is used for the analysis
of polar compounds. Chromosorb G, manufactured similar to the method used for
filter-aid supports, is considerably more durable. Its maximum loading is 5 wt%.

Figure 2.2   Scanning electron photomicrographs of (*a*) nonuniform coating of 20% Carbowax 20M on Chromosorb W-HP, 80/100 mesh (note the stationary phase "pooling" in the left-hand portion of the photograph), and (*b*) a more uniform coating of Carbowax 20M. (From ref. 1.)

It has been well established that the surfaces of diatomites are covered with silanol (Si–OH) and siloxane (Si–O–Si) groups. Pink diatomite is more adsorptive than white; this difference is due to the greater surface area per unit volume rather than to any fundamental surface characteristic. Pink diatomite is slightly acidic (pH 6 to 7), whereas white diatomite is slightly basic (pH 8 to 10). Both types of diatomites have two sites for adsorption: van der Waals sites and hydrogen-bonding sites. Hydrogen-bonding sites are more important, and there are two different types for hydrogen bonding: *silanol groups*, which act as a proton donor, and the *siloxane group*, where the group acts as a proton acceptor. Thus, samples containing hydrogen bonds (e.g., water, alcohol, amines) may show considerable tailing, whereas compounds that hydrogen-bond to a lesser degree (e.g., ketones, esters) do not tail as much.

A support should ideally be inert and not interact with sample components in any way; otherwise, a component may decompose on the column, resulting in peak tailing or even disappearance of the peak in a chromatogram. Active silanol groups (Si–OH functionalities) and metal ions constitute two types of active adsorptive sites on support materials. Polar analytes, acting as Lewis bases, can participate in hydrogen bonding with silanol sites and display peak tailing. The degree of tailing increases in the sequence hydrocarbons, ethers, esters, alcohols, carboxylic acids, and so on, and also increases with decreasing concentration of polar analyte. Treatment of the support with the most popular silylating reagent, dimethyldichlorosilane (DMDCS), converts silanol sites into silyl ether functionalities and generates a deactivated surface texture. Since this procedure is both critical and difficult (HCl is a product of the reaction), it is advisable to purchase DMDCS-deactivated support materials or column packings prepared with this chemically modified support material from a column manufacturer. *A word of caution:* The presence of moisture in a chromatographic system, due to either impure carrier gas or water content in injected samples, can hydrolyze silanized supports, reactivate them, and initiate degradation of many liquid phases.

Metal ions (e.g., Fe^{3+}) present on a diatomite support can cause similar decomposition of both sample and stationary liquid phase. These ions, which can be considered Lewis acids, can also induce peak tailing of electron-dense analytes (e.g., aromatics). These ions can be leached from the support surface by washing with hydrochloric acid followed by thorough rinsing to neutrality with deionized water of high quality. A Chromosorb support subjected to this treatment has the suffix AW after its name; the untreated or non–acid washed version of the same support is designated by the suffix NAW. A support that is both acid-washed and deactivated with DMDCS is represented as AW-DMDCS. The designation HP is used for the classification of a support as high-performance grade (i.e., the best quality available. Chromosorb supports and the popular Gas Chrom series of supports are listed in Table 2.2 as a function of type of diatomite and treatment, and pertinent support properties are displayed in Table 2.3.

It has become the practice to refer to particle sizing of chromatographic supports in terms of the mesh range. For sieving of particles for chromatographic columns, both Tyler Standard Screens and the U.S. Standard Series are frequently used. Tyler screens are identified by the actual number of meshes (openings) per linear inch. U.S. sieves are identified either by micrometer (micron) designations or by arbitrary numbers. A material referred to as 60/80 mesh will pass particles that through a 60-mesh screen but *Not* through an 80-mesh screen. You may also see this written as −60+80 mesh. Particle size is much better expressed in micrometers

TABLE 2.2 Selected Solid Supports[a]

Source	Acid-Washed, DMDCS-Treated	Non–Acid Washed	Acid-Washed
Firebrick	Chromosorb P-AW-DMDCS	Chromosorb P-NAW	Chromosorb P-AW
	Gas Chrom RZ	Gas Chrom R	Gas Chrom RA
Celite filter aid	Chromosorb W-AW-DMDCS Chromosorb W-HP[b]	Chromosorb W-NAW	Chromosorb W-AW
	Chromosorb G-AW-DMDCS Supelcoport[b]	Chromosorb G-NAW	Chromosorb G-AW
Other filter aid	Gas Chrom QII[b] Gas Chrom Q (also base washed, then silanized)		
	Gas Chrom Z	Gas Chrom S	Gas Chrom A

[a]Chromosorb, Gas Chrom, and Supelcoport are trademarks of Johns-Manville, Alltech, and Supelco, respectively.
[b]High-performance support or best available grade of support.

TABLE 2.3 Properties of Selected Chromosorb Diatomaceous Earth Supports

Support	Packing Density (g/mL)	Surface Area (m²/g)	Pore Volume (mL/g)	Maximum Liquid Phase Loading (%)
P-NAW	0.32–0.38	4–6	1.60	30
P-AW	0.32–0.38	4–6		
P-AW-DMDCS	0.32–0.38	4–6		
W-AW	0.21–0.27	1.0–3.5	3.56	15
W-HP	0.23	0.6–1.3		
G-NAW	0.49	0.5	0.92	5
G-AW-DMDCS	0.49	0.5		
G-HP	0.49	0.4		

Source: Data obtained from refs. 2 and 3.

(microns); therefore, 60/80 mesh would correspond to a particle size range of 250 to 177 μm. Table 2.4 shows the conversion of column-packing particle sizes and also the relationship between mesh size, micrometers, millimeters, and inches. Table 2.5 shows the relationship between particle size and sieve size. Also, it is customary for the particle size distribution of support materials designated by column manufacturers and in analytical methods to be expressed as a mesh range (e.g., 80 to 100 mesh).

Lack of the proper amount of packing in a gas chromatographic column often is the source of poor separation. How can one tell when a column is properly packed? The answer is twofold: by column performance (efficiency, i.e., number of theoretical plates) and by the peak symmetry (has a Gaussian or normal distribution shape). Many factors affect column performance; loosely packed columns generally are inefficient and are easily noticeable in glass columns. A column that is too

TABLE 2.4   Conversion Table of Column Packing Particles of Chromatographic Significance

Mesh Size	Micrometers	Millimeters	Inches
20	840	0.84	0.0328
30	590	0.59	0.0232
40	420	0.42	0.0164
50	297	0.29	0.0116
60	250	0.25	0.0097
70	210	0.21	0.0082
80	177	0.17	0.0069
100	149	0.14	0.0058
140	105	0.10	0.0041
200	74	0.07	0.0069
230	62	0.06	0.0024
270	53	0.05	0.0021
325	44	0.04	0.0017
400	37	0.03	0.0015
625	20	0.02	0.0008
1250	10	0.01	0.0004
2500	5	0.005	0.0002

TABLE 2.5   Relationship of Particle Size to Screen Openings

Sieve Size	Top Screen Openings (μm)	Bottom Screen Openings (μm)	Micrometer Spread
10/30	2000	590	1410
30/60	590	250	340
35/80	500	177	323
45/60	350	250	100
60/80	250	177	73
80/100	177	149	28
100/120	149	125	24
120/140	125	105	20
100/140	147	105	42

tightly packed gives excessive pressure drops or may even become completely plugged because the support particles have been broken (fractured) and fines are present. These fines usually accumulate near the end of a column.

Small quantities of acids and bases may also be added to the stationary phase to cover or neutralize active sites on a solid support. They usually have the same acid–base properties of the species being analyzed and are referred to as *tail reducers*. Phosphoric acid–modified packings are effective for analyzing fatty acids and phenols; potassium hydroxide has been used with success for amines and other basic compounds.

An often overlooked parameter in the selection of a packed column is the packing density of the support material. Packing density can have a rather pronounced effect on retention data. The stationary phase is coated on a support on a weight percent basis, whereas the packing material is placed in the column on a volume basis. If the packing density of a support increases, the total amount of stationary phase in the column will increase, even if the loading percentage is constant. Packing density varies among support materials (Table 2.3) and may even vary from batch to batch for a given type of support. Consider the following scenario. Two packings are prepared, 10% Carbowax 20M on Chromosorb G-HP (density 0.49 g/mL) the other Carbowax 20M (10%) on Chromosorb W-HP (density 0.23 g/mL) and each is subsequently packed into glass columns of identical dimensions. The column containing the impregnated Chromosorb W-HP will contain approximately twice as much stationary phase as will the other column. Therefore, to generate meaningful retention data and compare separations, careful attention should be paid to the nature and properties of a support.

Teflon Supports.   Although diatomite supports are widely used support materials, analysis of corrosive or very polar substances require even more inertness from the support. Halocarbon supports offer enhanced inertness, and a variety have been tried, including Fluoropak-80, Kel-F, Teflon, and other fluorocarbon materials. However, Chromosorb T made from Teflon 6 powder is perhaps the best material available, because high column efficiencies can be obtained when it is coated with a stationary phase that has a high surface area, such as polyethylene glycol. Chromosorb T has a surface area of 7 to 8 m^2/g, a packing density of 0.42 g/mL, an upper coating limit of 20%, and a rather low upper temperature limit of 250°C. Applications where this type of support are recommended are in analyses of water, acids, amines, HF, HCl, chlorosilanes, sulfur dioxide, and hydrazine. Difficulties in coating Chromosorb T and packing columns may be encountered, as the material tends to develop static charges. This situation is minimized by (1) using plasticware in place of glass beakers, funnels, and so on; (2) chilling the support to 10°C prior to coating; and (3) chilling the column before packing. References 4 to 8 provide further information for successful results with this support. However, preparation of columns containing Teflon-coated stationary phases is best performed by column manufacturers. The interested reader desiring further details on solid supports is urged to consult the comprehensive reviews of Ottenstein (9,10).

USP Designations of Solid Supports.   The validation of pharmaceutical methods is regulated by the *United States Pharmacopoeia* (USP), which lists solid Supports required for its various methods. The support designations for various VSP gas chromatographic methods are shown in Table 2.6. Also listed in this table are equivalent solid supports offered by other chromatographic suppliers. The USP-designated liquid phases and columns for USP gas chromatographic methods are presented later in the chapter.

Adsorbents for Gas–Solid Chromatography

Surface adsorption is the prevailing separation mechanism in gas–solid chromatography (GSC), where as great care is taken to avoid this effect in gas–liquid

TABLE 2.6   USP Designations of Popular Supports and Adsorbents

USP Nomenclature Code	USP Support Description	Similar Supports
S1A	Siliceous earth; has been flux-calcined with sodium carbonate flux and calcining above 900°C. The support is acid-washed, then water-washed until neutral, *but* not base-washed. The siliceous earth may be silanized by treatment with an agent such as dimethyldichlorosilane to mask surface silanol groups.	Silcoport, Chromosorb W-AW, Chromsorb W-HP, Supelcoport
S1AB	Siliceous earth treated as S1A and both acid- and base-washed.	Silcoport WBW, Supelcoport BW
S1C	A support prepared from crushed firebrick and calcined or burned with a clay binder above 900°C with subsequent acid-wash. It may be silanized.	Chromosorb PAW or PAW DMDCS
S1NS	Untreated siliceous earth.	Chromosorb W-NAW
S2	Styrene–divinylbenzene copolymer that has a nominal surface area of <50 m^2/g and an average pore diameter of 0.3 to 0.4 μm.	Chromosorb 101
S3	Styrene–divinylbenzene copolymer with a nominal surface area of 500 to 600 m^2/g and an average pore diameter of 0.0075 μm.	Hayesep Q, Porapak Q, Super Q
S4	Styrene–divinylbenzene copolymer with aromatic–O and–N groups that has a nominal surface area of 400 to 600 m^2/g and an average pore diameter of 0.0076 μm.	Hayesep R, Porapak R
S5	40- to 60- mesh high-molecular-weight tetrafluoroethylene polymer.	Chromosorb T
S6	Styrene–divinylbenzene copolymer that has a nominal surface area 250 to 350 m^2/g and an average pore diameter of 0.0091 μm.	Chromosorb 102, Hayesep P, Porapak P
S7	Graphitized carbon that has a nominal surface area of 12 m^2/g.	Carbopack C, CarboBlack C
S8	Copolymer of 4-vinylpyridine and styrene–divinylbenzene.	Hayesep S, Porapak S
S9	Porous polymer based on 2,6-diphenyl-p-phenylene oxide	Tenax TA
S10	Highly polar cross-linked copolymer of acrylonitrile and divinylbenzene.	Hayesep C
S11	Graphitized carbon that has a nominal surface area of 100 m^2/g modified with small amounts of petrolatum and polyethylene glycol compound.	3% SP-1500 on 80/120-mesh Carbopack B, CarboBlack B 80/120-mesh 3% Rt 1500
S12	Graphitized carbon that has a nominal surface area of 100 m^2/g.	Carbopack B, CarboBlack B

Source: *USP Column Cross-Reference Charts:* Restek Corporation and Supelco.

chromatography (GLC). In GSC an uncoated adsorbent serves as the column packing, although special effects in selectivity by a mixed retention mechanism can be obtained by coating the adsorbent with a stationary phase. The latter case is an illustration of gas–liquid–solid chromatography (GLSC). Permanent gases and very volatile organic compounds can be analyzed by GSC, as their volatility, which causes rapid elution from the column, is problematic in GLC. Another fertile field is the determination of adsorption thermodynamic functions by GSC. The most exhaustive work in gas–solid chromatography had been with charcoal, alumina, and silica gel.

Inorganic Salts.   An initial investigation into the use of inorganic salts as stationary phases (adsorbents) for GSC was the work of Hanneman (11). This study employed a eutectic mixture (15°C below its melting point) of lithium, sodium, and potassium nitrates. Favre and Kallenbach (12) studied the separation of *o-*, *m-*, and *p*-terphenyls on columns containing inorganic salts (25% wt/wt) adsorbed on Chromosorb P. They found that little relationship existed between cation–anion changes and suggested that adsorption was dependent primarily on surface characteristics. Later, Salomon (13) studied *o-*, *m-*, and *p*-terphenyls and a variety of other compounds displaying various polarities as sorbates. His column adsorbents were alkali metal chlorides, sulfates, and carbonates on a Chromosorb P support. He concluded that more-polar compounds eluted at a higher temperature than did less-polar compounds, although the boiling points may be the same. Salomon's conclusions were as follows:

1. The separations were affected by weak bonding between the inorganic salt and the organic molecules.
2. The separations were affected by a modification of the adsorptive sites of the column support, which may then react with the organic molecules.
3. The melting points of the sorbent solids had little effect on column performance.
4. If two inorganic salts are mixed (as adsorbents), retention data comprised the average of the two when adjusted for concentration.
5. The anion–cation effects of the adsorbents are significant. For example, sulfates cause elution temperature to increase compared to chlorides.

Rogers and Altenau (14–16) studied inorganic complexes, which displayed an array of adsorptive properties. They found that water, pyridine, or ammonia could be pyrolytically eliminated to produce very porous solids with relatively large surface areas compared to those of the starting materials. Their studies showed that oxygenated sorbates were more strongly adsorbed than aliphatic sorbates. Amines, with a lone pair of electrons, did not desorb. They concluded that interaction occurs between the metal in the complex and either the lone pair or pairs of electrons in nitrogen or oxygen or with a π electron system in aromatic compounds, and finally, by induced dipoles in the aliphatic compounds.

Grob et al. (17) studied the alkali metal (Li, Na, K, Rb, Cs) chlorides and nitrates as possible adsorbents for GSC. Each salt was sieved to 60/80 mesh and packed into metal columns (320 cm × 0.23 in. i.d.). This study showed the following:

1. The alkali metal nitrate columns had a greater retention order than that of the corresponding alkali metal chloride columns.
2. The retention of sorbates generally followed the boiling point of sorbates. For two compounds that had the same boiling point, the more-polar compound had the greater retention time.
3. Chain branching reduced retention through decreased polarity.
4. Electron-releasing groups enhanced adsorption; electron-withdrawing groups decreased adsorption.
5. Ortho substituents caused stronger adsorptive effects than those of meta substituents.
6. The adsorption of solutes is due to physical adsorption.
7. Heats of adsorption increase as the polarity or electron densities of the sorbates increase.

Continued investigations by Sawyer et al. (18–22) showed specific interactions affecting retention, thermodynamics and separation efficiencies, and correlations of aromatic substituent effects. These studies led to a thermodynamic gas chromatographic retention index for organic molecules.

Because of the great potential selectivity of solids, GSC is especially desirable for the separation of isomers and high-molecular-weight compounds which are thermally stable. These properties of GSC prompted research into the use of barium and strontium halides (chloride, bromide, and iodide) as adsorbent supports for GSC (all columns were made of glass; 6 ft × 0.25 in. o.d.) (23). Conclusions from this study showed the following:

1. Retention volumes generally followed boiling points.
2. Two compounds with similar boiling points showed greater adjusted retention volume for the more-polar compound.
3. Chain branching reduced retention volume but increased heats of adsorption.
4. Retention volumes increased as the size of anion in the solid support increased.
5. Electron-releasing groups (e.g., alkyl groups) on benzene rings cause increases in heats of adsorption due to an increase in the π electron density in the ring.
6. Electron-withdrawing groups (e.g., halogen groups) on benzene rings decrease adsorption (decrease in the π electron density in the ring).

Studies into the use of GSC began to take on more depth. These resulted from earlier studies into physical adsorption and chemisorption studies prior to chromatographic investigations. Garner and Veal (24) and Garner (25) had defined three processes, each of which could be termed *adsorption*. The first of these was *physical adsorption*, due primarily to van der Waals forces. The latter two were *reversible* and *irreversible chemisorption*, which could be attributed to bond formation between the adsorbent and the sorbate. Kipling and Peakall (26) observed that no single experimental measurement could provide a criterion to distinguish between chemisorption and physical adsorption. Their observation was that there were many exceptions to the general concept that the heat of sorption for physical adsorption is less than that for chemisorption. They concluded that in a chemisorption process all three operations generally occur and that the determined heats of sorption reflect the average contribution of all three processes.

Dent and Kohes (27) proposed a definite bond between propylene and zinc oxide, infrared spectroscopy being used to study the sorption process. A series of articles by Gil-Av et al. (28–31) demonstrated the use of stationary phases containing silver nitrate to separate saturated and unsaturated compounds.

As a follow-up, Grob and McGonigle (32,33) coupled the column selectivity and high-temperature operation properties of the chlorides of vanadium(II), manganese(II), and cobalt(II) for the separation of various unsaturated compounds. These inorganic packings differ primarily in their number of available $3d$ electrons: V(II) ($3d^3$), Mn(II) ($3d^5$), and Co(II) ($3d^7$). This study was conducted to observe the influence of the various π-electron densities. All the inorganic packed columns were 6 ft $\times$ 0.25 in. o.d. glass columns.

Conclusions from the investigation revealed the following:

1. Heats of adsorption, $-\Delta H_a$, varied directly with the π-electron density of the adsorbates and inversely with the number of $3d$ electrons of the inorganic packings (sorbents).
2. Isolated π bonds in the adsorbates were better for separations than compounds with conjugated π bonds.
3. Electronegative groups close to the π bonds in the adsorbate increased adsorption.
4. Heats of adsorption, $-\Delta H_a$, increase in the order $3d^5$, $3d^7$, $3d^3$ (i.e., MnCl$_2$, CoCl$_2$, VCl$_2$).
5. Temperatures greater than twice the boiling point of the adsorbates were necessary to elute unsaturated compounds from VCl$_2$.
6. Transition metal chloride packings offer a practical basis for quantitative separation of linear isomeric hydrocarbons with increasing degrees of unsaturation (LIHIDUs) and cyclic isomeric hydrocarbons with increasing degrees of unsaturation (CIHIDUs).

A series of metal oxides have also been investigated as possible packings for GSC (34): V_2O_3, V_2O_4, V_2O_5, SnO_2, TiO, and ZnO. These metal oxides are also catalysts. Chromatographic evidence of specific interactions was substantiated using attenuated total reflectance (ATR) in the infrared (IR) region. This type of study is well adapted to surface studies and enables one to use the column packing as the reflecting surface (34,35). In this study, column packing containing the sorbate was placed on both sides of the KRS-5 crystal (an ATR crystal composed of Tl^+, Br^-, I^-). An ATR-IR spectrum scan was made of the sorbate and compared to an IR spectrum scan of the "neat" sorbate. The ATR-IR spectrum scan of the sorbate was distinctly different from the IR spectrum scan of the sorbate, indicating an interaction between the inorganic packing and the sorbate(s).

Thermodynamics provided the theoretical and experimental framework for judging the characteristics of the sorbates and generalizations concerning the nature of the interactions observed. For example, van't Hoff plots (ln V_g versus $1/T$) were employed to calculate the heats of adsorption ($-\Delta H_{ads}$) of the individual sorbates on the various inorganic packings(sorbents). In the practice of GSC, another measurement of the sorbent is its surface area. A number of techniques are available to obtain measurements of the surface area (36,37).

The use of inorganic salt supports was extended after the invention of capillary columns, or *open-tubular columns* as they are referred to, by the investigations of Marcel Golay (38) and later improved by the work of Dandeneau and Zerenner (39,40) with the introduction of fused silica as the column material. Before the commercial availability of fused-silica open-tubular columns, chromatographers made their own glass capillary columns by means of glass-drawing machines (e.g., Model GDM-1 glass drawing machine, Shimadzu Scientific Instruments, Columbia Maryland). Wishousky et al. (41,42) investigated whisker-walled open-tubular (WWOT) columns coated with $MnCl_2$ and $CoCl_2$. The main purpose of these studies was twofold: to compare packed columns to capillary columns and to study the advantage of increased surface area of the inorganic salt packings when deposited on the whiskered walls of the capillary columns. 2-Chloro-1,1,2-trifluoroethylmethyl ether was used to etch the glass columns and grow the whiskers, and octamethylcyclotetrasiloxane (OMCTS) was used to deactivate such columns. These inorganic salt-coated capillary columns proved useful to study the interactions of aromatic hydrocarbons that possessed varying π-electron densities and alkyl substitutions. Halogenated aromatic hydrocarbons eluted according to boiling points, whereas the retention of alkyl-substituted benzenes increased with alkyl chain length and degree of substitution. These capillary columns provided increased performance over that of packed columns using the same inorganic salts. The dynamic coating procedure was more practical than the static coating procedure for the preparation of these whisker-walled columns. The optimum length of these columns was shown to be 25 to 30 m.

The use of columns packed with inorganic salts was an active area of gas chromatography in the late 1970s and early 1980s. However, the advent of chemically bonded phases (for capillary columns) increased in interest, ease of preparation, and stability, and inorganic salt packings lost the interest of gas chromatographers. It was, however, an era of interesting research, especially for the study of mechanisms of separations and the gathering of thermodynamic data.

Porous Polymers. Porous polymers are the adsorbents of choice for most applications focusing on the analysis of gases, organics of low carbon number, acids, amines, and water (43,44). The presence of water is detrimental to gas–liquid chromatographic packings. Because water is eluted with symmetrical band profiles on a number of porous polymers, these adsorbents may be employed for the analyses of aqueous solutions and the determination of water in organic matrices. There are three separate product lines of commercially available porous polymers: the Porapaks (Millpore Corp.), the Chromosorb century series (Johns-Manville), and HayeSep (Hayes Separation) polymers. Within each product line there are several members, each differing in chemical composition and therefore exhibiting unique selectivity, as shown in Table 2.7. On the other hand, some adsorbents are quite similar, as in the case of Porapak Q and Chromosorb 102 (both styrene–divinylbenzene copolymers) and HayeSep C and Chromosorb 104 (both acrylonitrile–divinylbenzene copolymers). In the future we should expect to see new polymers that address old separation problems, as was the case with the arrival of HayeSep A, which can resolve a mixture of nitrogen, oxygen, argon, and carbon monoxide at room temperature (45).

Molecular Sieves.   These sorbents, also referred to as *zeolites*, are synthetic alkali or alkaline earth metal aluminum silicates employed for the separation of hydrogen, oxygen, nitrogen, methane, and carbon monoxide. These gases are separated on molecular sieves because the pore size of the sieve matches the molecular diameters of the gases. Two popular types of molecular sieves are used in GSC; Molecular Sieve 5A (pore size of 5 Å with calcium as the primary cation) and Molecular Sieve 13X (pore size of 13 Å with sodium as the primary cation). At normal temperatures, molecular sieves permanently adsorb carbon dioxide, which gradually degrades the O_2–N_2 resolution. The use of a silica gel precolumn that adsorbs carbon dioxide eliminates this problem. Molecular sieve columns must be conditioned at $300°C$ to remove residual moisture from the packing; otherwise, the permanent gases elute too quickly, with little or no resolution, and coelution or reversal in elution order for CO–CH_4 may occur (46).

Carbonaceous Materials.   Adsorbents containing carbon are available commercially in two forms: carbon molecular sieves and graphitized carbon blacks. The use of carbon molecular sieves as packings for GSC was first reported by Kaiser (47). They behave similarly to molecular sieves because their pore network is also in the angstrom range. Permanent gases and C_1–C_3 hydrocarbons may be separated on carbonaceous sieves such as Carbosphere and Carboxen.

Graphitized carbons play a dual role in GC. They are a nonspecific adsorbent in GSC with a surface area in the range 10 to 1200 m^2/g. Adsorbents such as Carbopacks and Graphpacs may also serve as a support in GLC and in GLSC, where unique selectivity is acquired and a separation is based on molecular geometry and polarizability considerations. Coated graphitized carbons can tolerate aqueous samples and have been used by DiCorcia and co-workers for the determination of water in glycols, acids, and amines (48–50). In the latter roles, since graphitized carbon has a nonpolar surface, it must be coated with a stationary phase for deactivation of its surface. The resulting packing reflects a separation that is a hybrid

TABLE 2.7   Porous Polymeric Adsorbents for Gas–Solid Chromatography

Adsorbent	Polymeric Composition or Polar Monomer (PM)[a]	Maximum Temperature (°C)	Applications
HayeSep A	DVB–EGDMA	165	Permanent gases, including hydrogen, nitrogen, oxygen, argon, CO, and NO at ambient temperature; can separate C2 hydrocarbons, hydrogen sulfide, and water at elevated temperatures
HayeSep B	DVB–PEI	190	C1 and C2 amines; trace amounts of ammonia and water.
HayeSep C	ACN–DVB	250	Analysis of polar gases (HCN, ammonia, hydrogen sulfide) and water.
HayeSep D	High-purity DVB	290	Separation of CO and carbon dioxide from room air at ambient temperature; elutes acetylene before other C2 hydrocarbons; analyses of water and hydrogen sulfide.
Porapak N	DVB–EVB–EGDMA	190	Separation of ammonia, carbon dioxide, water, and separation of acetylene from other C2 hydrocarbons.
HayeSep N	EGDMA (copolymer)	190	
Porapak P	Styrene–DVB	250	Separation of a wide variety of alcohols, glycols, and carbonyl analytes.
HayeSep P	Styrene–DVB	250	
Porapak Q	EVB–DVB copolymer	250	Most widely used; separation of hydrocarbons, organic analytes in water, and oxides of nitrogen.
HayeSep Q	DVB polymer	275	
Porapak R	Vinyl pyrollidone (PM)	250	Separation of ethers and esters; separation of water from chlorine and HCl.
HayeSep R	250		
Porapak S	Vinyl pyridine (PM)	250	Separation of normal and branched alcohols.
HayeSep S	DVB–4-vinyl pyridine	250	
Porapak T	EGDMA (PM)	190	Highest-polarity Porapak; offers greatest water retention; determination of formaldehyde in water.
HayeSep T	EGDMA polymer	165	

(Continued)

TABLE 2.7   (*continued*)

Adsorbent	Polymeric Composition or Polar Monomer (PM)[a]	Maximum Temperature (°C)	Applications
Chromosorb 101	Styrene–DVB	275	Separation of fatty acids, alcohols, glycols, esters, ketones, aldehydes, ethers, and hydrocarbons.
102	Styrene–DVB	250	Separation of volatile organics and permanent gases; no peak tailing for water and alcohols.
103	Cross-linked PS	275	Separation of basic compounds, such as amines and ammonia; useful for separation of amides, hydrazines, alcohols, aldehydes, and ketones.
104	ACN–DVB	250	Nitriles, nitroparaffins, hydrogen sulfide, ammonia, sulfur dioxide, carbon dioxide, vinylidene chloride, vinyl chloride, and trace water content in solvents.
105	Cross-linked polyaromatic	250	Separation of aqueous solutions of formaldehyde; separation of acetylene from lower hydrocarbons and various classes of organics with boiling points up to 200°C.
106	Cross-linked PS	225	Separation of C2 to C5 alcohols; separation of C2 to C5 fatty acids from corresponding alcohols.
107	Cross-linked acrylic ester	225	Analysis of formaldehyde, sulfur gases, and various classes of compounds.
108	Cross-linked acrylic	225	Separation of gases and polar species such as water, alcohols, aldehydes, ketones, glycols, etc.

Source: Data from refs. 8, 15, and 16.
[a]DVB, divinyl benzene; EGDMA, ethylene glycol dimethacrylate; PEI, polyethyleneimine; ACN, acrylonitrile; EVB, ethylvinyl benzene.

of gas–solid and gas–liquid mechanisms. Frequently, the packing is modified further by the addition of phosphoric acid (H_3PO_4) or potassium hydroxide (KOH) to reduce peak tailing for acidic and basic compounds, respectively. Separations of alcohols and amines are displayed in Figure 2.3. The USP support designations specified in many gas chromatographic methods appear in Table 2.6.

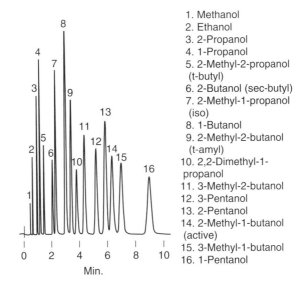

1. Methanol
2. Ethanol
3. 2-Propanol
4. 1-Propanol
5. 2-Methyl-2-propanol (t-butyl)
6. 2-Butanol (sec-butyl)
7. 2-Methyl-1-propanol (iso)
8. 1-Butanol
9. 2-Methyl-2-butanol (t-amyl)
10. 2,2-Dimethyl-1-propanol
11. 3-Methyl-2-butanol
12. 3-Pentanol
13. 2-Pentanol
14. 2-Methyl-1-butanol (active)
15. 3-Methyl-1-butanol
16. 1-Pentanol

80/100 Carbopack C/0.2% Carbowax 1500, 6′ × 2mm ID glass, Col. Temp.: 135°C, Flow Rate: 20mL/min., N_2, Det.: FID, Sample Size: 0.02μL

(a)

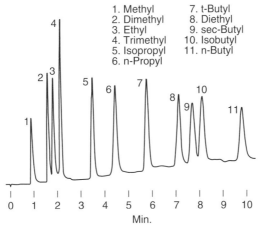

1. Methyl
2. Dimethyl
3. Ethyl
4. Trimethyl
5. Isopropyl
6. n-Propyl
7. t-Butyl
8. Diethyl
9. sec-Butyl
10. Isobutyl
11. n-Butyl

60/80 Carbopack B/4% Carbowax 20M/0.8% KOH, 6′ × 2mm ID glass, Col. Temp.: 90°C to 150°C @ 4°C/min., Flow Rate: 20mL/min., N_2, Det.: FID, Sens.: 4 × 10^{-11} AFS, Sample: 0.5μL, 100ppm each amine in water.

(b)

Figure 2.3 Separation of C_1 to C_5 alcohols (a) and aliphatic amines (b) on graphitized carbon. (From ref. 46; copyright © 1985 John Wiley & Sons, Inc., with permission.)

2.3   STATIONARY PHASES

Requirements of a Stationary Phase

An ideal stationary liquid phase for GLC should exhibit selectivity and differential solubility of the components to be separated and a wide range of operating temperature. A phase should be chemically stable and have a low vapor pressure at elevated column temperatures ($\ll 0.1$ torr). A minimum temperature limit near ambient temperature, where the liquid phase still exists as a liquid and not as a solid, is desirable for separations at or near room temperature and eliminates a gas–solid adsorption mechanism (e.g., Carbowax 20M is a solid at temperatures below $60°$C. In choosing a liquid phase, some fundamental criteria must be considered:

- Is the liquid phase selective toward the components to be separated?
- Will there be any irreversible reactions between the liquid phase and the components of the mixture to be separated?
- Does the liquid phase have a low vapor pressure at the operating ($\ll 0.1$ torr) temperature? Is it thermally stable?

Let us consider some of the information available to answer these questions. We are not now attempting to develop a pattern for selection of the proper liquid substrate; this is discussed in Chapter 3. The vapor pressure of the liquid phase should be less than 0.1 torr at the operating temperature, as mentioned above. This value can change depending on the detector used, since bleed from the liquid phase will cause noise and elevate background signal and thus decrease sensitivity. Information from a plot of vapor pressure versus temperature is not always completely informative or practical because adsorption of the liquid phase on the solid support results in a decrease in the actual vapor pressure of the liquid phase. Other than its effect on detector noise, liquid-phase bleed may interfere with analytical results and determine the life of the column. Also, some supports may have a catalytic effect, causing decomposition of the liquid phase, thereby reducing its life in the column. Contaminants in the carrier gas (e.g., O_2) may also cause premature fatigue of a liquid phase. The effect of impurities in the various gases used in gas chromatography has been well addressed by Bartram (51). Two other properties of the liquid phase to be considered are viscosity and wetting ability. Ideally, liquid phases should have low viscosity and high wetting ability (i.e., the ability to form a uniform film on the solid support or column wall).

It is uninformative to refer to a liquid phase as being selective, since all liquid phases are selective to varying degrees. *Selectivity* refers to the relative retention or band spacing of two components and gives no information regarding the mechanism of separation. Most separations depend on the boiling-point difference, variation in molecular weights of the components, and the structure of the components being separated.

The *relative volatility* or *separation factor*, α, depends on the interactions of the solute and the liquid phase: that is, van der Waals cohesive forces, which may be divided into three types:

1. *London dispersion forces*: caused by the attraction of dipoles that arise from the arrangement of the elementary charges. Dispersion forces act between all molecular types, especially in the separation of nonpolar substances (e.g., saturated hydrocarbons).
2. *Debye induction forces*: result from interaction between permanent and induced dipoles.
3. *Keesom orientation forces*: result from the interaction of two permanent dipoles, of which the hydrogen bond is more important. Hydrogen bonds are stronger than dispersion forces. If the two components have the same vapor pressure, separation can be achieved on the basis of several properties: (1) difference in the functional groups > (2) isomers with polar functional groups > (3) isomers with no functional groups.

In the selection of a stationary phase, a compromise must be struck between theory and practice. For example, theory dictates that a stationary phase of low viscosity or fluid in texture is preferable to a chemically equivalent, more viscous gum phase, as may be ascertained from the contribution of D_l, the solute diffusivity in the stationary phase in equation 2.1. However, this same fluid, possessing a lower-molecular-weight or weight distribution, if polymeric in nature, will typically have poorer thermal stability and a lower maximum operating temperature. Although unfavorable from the viewpoint of mass transfer in the van Deemter expression in equation 2.1, practical considerations may favor the gum for separations requiring high column temperatures. Equation 2.1 indicates that higher column efficiency is obtained with a column containing a low percentage loading of stationary phase than with the same column packed with a higher phase loading. But in practice, deposition of a thin coating of stationary phase on a support may yield insufficient coverage of the active sites on the surface of the support, resulting in peak tailing and reestablish a need for a higher percentage of stationary-phase loading. Note in Figure 2.4b the peak tailing of the n-alkanes on a lightly loaded packing (<3% OV-101 on Chromosorb W-HP) and the elimination of tailing with a heavier coating of stationary phase (Figure 2.4a).

Separations in GLC are the result of selective solute–stationary phase interactions and differences in the vapor pressure of solutes. The main forces that are responsible for solute interaction with a stationary phase are dispersion, induction, orientation, and donor–acceptor interactions (52–54), the sum of which serves as a measure of the *polarity* of the stationary phase toward the solute. *Selectivity*, on the other hand, may be viewed in terms of the magnitude of the individual energies of interaction. In GLC, the selectivity of a column governs band spacing, the degree to which peak maxima are separated. The following parameters influence selectivity:

1. Nature of the stationary phase
2. Percent loading of the stationary phase

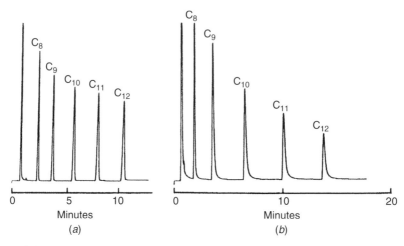

Figure 2.4   Chromatograms of *n*-alkanes on (*a*) 6-ft glass column, 2 mm i.d. containing 20% OV-101 on 80/100-mesh Chromosorb W-HP, column conditions 100 to 175°C at 6°C/min; (*b*) same as in part (*a*) but with 3% OV-101 on the same support, column conditions 50 to 120°C at 4°C/min. Flow rate 25 mL/min He, detector FID. (From ref. 1.)

3. Column temperature
4. Choice and pretreatment of solid support or adsorbent

Differences in selectivity are significant because they permit the separation of solutes of similar or even the same polarity by a selective stationary phase.

In the early practice of gas chromatography, the concept of polarity and even the requirements for a stationary phase were not clearly understood. There was a proliferation of liquid phases, encompassing (1) those with marginal gas chromatographic properties, such as Nujol, glycerol, diglycerol, and even Tide (the laundry detergent); (2) industrial-grade lubricants of variable composition, such as the Apiezon greases and Ucon oils; and (3) an abundance of phases that just simply duplicated the chromatographic behavior of others. In retrospect, the vast array of stationary phases can be attributed to the fact that mechanisms of separation were not well understood, and many "chromatographers" tried just about any chemical they had on the laboratory shelf or the organic stockroom. This hit-and-miss search for a unique stationary phase was pursued to compensate for the inefficiency of a packed column coupled with the search for some degree of selectivity for the resolution of two solutes [equation 2.9]. The development of capillary columns with their high efficiency, and a clearer understanding of column selectivity permitted the chromatographer to decrease the number of stationary phases, each differing in selectivity, to achieve any required resolution.

The stationary-phase requirements of selectivity and higher thermal stability then became more clearly defined; the process of stationary-phase selection and classification became logical after the studies of McReynolds (55) and Rohrschneider (56, 57) were published, both of which were based on the retention index (58). The

Kovats' retention index procedure and the McReynolds and Rohrschneider constants are discussed in detail in the following sections. The Kovats index remains a widely used technique for reporting retention data, and every stationary phase developed for packed and capillary gas chromatography has been characterized by its McReynolds constants.

USP Designation of Stationary Phases

The stationary phases designated for the USP methods are listed in Table 2.8. Also listed are equivalent stationary phases recommended by other chromatographic suppliers.

Kovats Retention Index

The Kovats approach solved problems pertaining to the use, comparison, and characterization of gas chromatographic data. The reporting of retention data as *absolute retention time, t_R,* is meaningless because virtually every chromatographic parameter and any related experimental fluctuation affect a retention-time measurement. The use of *relative retention data* $(\alpha = t'_{R2}/t'_{R1})$ offered some improvement, but the lack of a universal standard suitable for wide temperature ranges on stationary phases of different polarities has discouraged its utilization. In the Kovats approach, the *retention index, I,* of an *n*-alkane is assigned a value equal to 100 times its carbon number. For example, the I values of *n*-octane, *n*-decane, and *n*-dodecane are equal to 800, 1000, and 1200, respectively, by definition and are applicable on any column, packed or capillary, any liquid phase, and independent of every chromatographic condition, including column temperature. However, for all other compounds, chromatographic conditions such as stationary phase and its concentration, support, and column temperature for packed columns must be specified. Since retention indices are also the preferred method for reporting retention data with capillary columns, the stationary phase film thickness, and column temperature also have to be specified for compounds other than *n*-alkanes; otherwise, the I values are meaningless.

An I value of a compound can be determined by spiking a mixture of *n*-alkanes with the component(s) of interest and chromatographing the resulting mixture under the conditions specified. A plot of log adjusted retention, t_R, versus the retention index, I, is generated, and the retention index of the solute under consideration is determined by extrapolation, as depicted in Figure 2.5 for isoamyl acetate. The selectivity of a particular stationary phase can be established by comparing the I values of a solute on a nonpolar phase such as squalane or OV-101 ($I = 872$) with a corresponding I value of 1128, associated with a more polar column containing Carbowax 20M, for example. This difference of 256 units indicates the greater retention produced by the Carbowax 20M column. More specifically, isoamyl acetate elutes between *n*-C_{11} and *n*-C_{12} on Carbowax 20M but more rapidly on OV-101, where it elutes after *n*-octane.

TABLE 2.8   USP Designation of Stationary Phases

USP Nomenclature Code	USP Stationary-Phase Description	Similar Stationary Phases
G1	Dimethylpolysiloxane oil	Rtx-1, Rtx-2100,SP-2100, OV-101 SE-30, Equity-l(capillary), SPB-1 (capillary), MDN-1(capillary), BP1, SolGel-1 ms, HP-1, DB-1, HP-1 ms, DB-1 ms
G2	Dimethylpolysiloxane gum	Rtx-1, SP-2100, OV-1, SE-30 Equity-1(capillary), SPB-1 (capillary), MDN-l(capillary), BP1, SolGel-1 ms, HP-1, DB-1, HP-1 ms, DB-1 ms
G3	50% Phenyl-and 50% methylpoly siloxane	Rtx-50, Rtx-2250, OV-17, SP-2250, SPB-50(capillary), SP-2250(capillary), SPB-17 (capillary), BPX50, DB-17, HP-50+
G4	Diethylene glycol succinate polyester	Rt-DEGS
G5	3-Cyanopropylpoly-siloxane	Rt-2340, SP-2340, Silar 10 CP, SP-2340(capillary), SP-2560 (capillary), BPX70, DB-23
G6	Trifluoropropyl-methylpolysiloxane	Rt-2401, Silicone OV-210, SP-2401, DB-23
G7	50% 3-Cyanopropyl-and 50% phenylmethylsilicone	Rtx-225, Rt-2300, SP-2300, Silar 5 CP,SPB-225(capillary), BP225,DB-225,DB-225 ms
G8	90% 3-Cyanopropyl-and 10% phenylmethylsilicone	Rtx-2330,SP-2330,SP-2330 (capillary)
G9	Methylvinylpoly-siloxane	UCW-98,OV-1,UC W982, Equity-1(capillary),SPB-1 (capillary),MDN-1(capillary)
G10	Polyamide formed by reacting a C36 dicarboxylic acid with 1,3,4-piperdylpropane and piperdine in the respective mole ratios of 1.00:0.90:0.20	Polyamide, Poly-A 103
G11	Bis(2-ethylhexyl)sebacate polyester	Di(2-ethylhexyl)sebacate
G12	Phenyldiethanolamine succinate polyester	Phenylethanolamine succinate
G13	Sorbitol	Sorbitol
G14	Polyethylene glycol(average mol. wt. of 950 to 1050)	Carbowax 1000,BP20(WAX), SolGel-WAX,DB-WAX
G15	Polyethylene glycol (average mol. wt. of 3000 to 3700)	Carbowax 4000,BP20(WAX), SolGel-WAX,DB-WAX

TABLE 2.8 (*continued*)

USP Nomenclature Code	USP Stationary-Phase Description	Similar Stationary Phases
G16	Polyenthylene glycol compound (average mol. wt. about 15,000); high-molecular-weight compound of polyethylene glycol with a diepoxide linkage	Stabilwax, Carbowax 20M, Omegawax(capillary), Supelcowax 10(capillary), BP20(WAX)SolGel-WAX, DB-WAX
G17	75% Phenyl-and 25% methylpolysi loxane	OV-25,BPX50,DB-17,HP-50+
G18	Polyalkylene glycol	Ucon LB-1800-X, UCON LB-550-X,PAG(capillary)
G19	25% Phenyl-,25% cyanopropyl-, and 50% methylsilicone	Rtx-225,OV-225,SPB-225 (capillary), BP225, DB-225, DB-225 ms
G20	Polyethylene glycol(average mol. wt. of 380 to 420)	Carbowax 400,BP20(WAX), SolGel-WAX, DB-WAX
G21	Neopentyl glycol succinate	Neopentyl glycol succinate
G22	Bis(2-ethylhexyl)phthalate	Bis(2-ethylhexyl) phthalate
G23	Polyethylene glycol adipate	EGA
G24	Diisodecyl phthalate	Diisodecyl phthalate
G25	Polyethylene glycol compound TPA; high-molecular-weight compound of a polyethylene glycol and a diepoxide that is esterfied with terephthalic acid.	Stabilwax-DA, Carbowax 20MTPA,SP-1000,free fatty acid phase(FFAP),SPB-1000 (capillary, Nukol(capillary), BP21(FFAP),DB-FFAP, HP-FFAP
G26	25% 2-Cyanoethyl-and 75% methyl polysiloxane	Rt-XE-60
G27	5% Phenyl-and 95% methylpolysiloxane	Rtx-5,SE-52, Equity-5(capillary), PTE-5(capillary), SPB-5(capillary), MDN-5(capillary), BP5,BPX5, DB-5,HP-5, HP-5 ms, DB-5 ms
G28	25% Phenyl-and 75% methylpolysiloxane	Rtx-20,DC-550, BPX35, DB-35, HP-35, DB-35 ms
G29	3,3'-Thiodipropionitrile	β, β'-Thiodipropionitrile(TDPN)
G30	Tetraethylene glycol dimethyl ether	Tetraethylene glycol dimethyl ether
G31	Nonylphenoxypoly(ethyl-eneoxy) ethanol(average ethyleneoxy chain length is 30); Nonoxynol 30	Igepal CO-880(nonoxynol)
G32	20% Phenylmethyl-80% dimethyl polysiloxane	Rtx-20, OV-7, SPB-20(capillary), BPX35,DB-35, HP-35,DB-35 ms
G33	20% Carborane−80% dimethyl polysiloxane	20% Carborane−80% dimethyl polysiloxane, Dexsil 300
G34	Diethylene glycol succinate polyester stabilized with H_3PO_4	Rt-DEGS PS, DEGS-PS

(Continued)

TABLE 2.8   (*continued*)

USP Nomenclature Code	USP Stationary-Phase Description	Similar Stationary Phases
G35	High-molecular-weight compound of polyethylene glycol and a diepoxide that is esterified with nitroterephthalic acid	Stabilwax-DA,Rt-1000,Carbowax 20M-TPA,SP-1000, freefatty acid phase(FFAP),SPB-1000 (capillary), Nukol(capillary), Carbowax 20M-terephthalic acid, BP21(FFAP), DB-FFAP, HP-FFAP
G36	1% Vinyl-and 5% phenylmethylpolysiloxane	Rtx-5,SE-54,Equity-5(capillary), PTE-5(capillary), SPB-5 (capillary), SE-54(capillary), MDN-5(capillary),BP5,BPX5, DB-5,HP-5,HP-5 ms, DB-5 ms
G37	Polyimide	Poly-I 110
G38	Phase G1 containing a small percentage of a tailing inhibitor	Rt-2100/0.1%Carbowax 1500, SP-2100 + 0.1% Carbowax 1500, SP-2100 + 0.2% Carbowax 1500, BP1, SolGel-1 ms, DB-1,HP-1, HP-1 ms, DB-1 ms
G39	Polyethylene glycol(average mol. wt. about 1500)	Carbowax 1500, BP20(WAX), SolGel-WAX,DB-WAX
G40	Ethylene glycol adipate	Rt-EGA
G41	Phenylmethyldimethyl-siloxane (10% phenyl-substituted)	Rtx-5,OV-3,BP5, BPX5, DB-5, HP-5, HP-5 ms,DB-5 ms
G42	35% Phenyl-and 65% dimethylpolysilox-ane(percentages refer to molar substitution)	Rtx-35,OV-11,SPB-35(capillary), BPX35,DB-35,HP-35,DB-35 ms
G43	6% Cyanopropylphenyl-and 94% dimethylpolysilox-ane(percentages refer to molar substitution)	Rtx-624,Rtx-1301, OV-1301, SPB-624(capillary),OVI-G43 (capillary), BP624, DB-624 DB-1301
G44	2% Low molecular-weight petrolatum hydrocarbon grease and 1% solution. of KOH	Apiezon L + 1% KOH
G45	Divinylbenzene–ethylene glycol–dimethylacrylate	Rt-QPLOT,HayeSep A,HayeSep N,Porapak N,HP-PLOT U
G46	14% Cyanopropylphenyl-and 86% methylpolysiloxane	Rtx-1701,OV-1701,Equity 1701 (capillary), SPB-1701(capillary), BP10(1701), DB-1701
G47	Polyethylene glycol(average mol. wt. of about 8000)	Carbowax 8000
G48	Highly polar, partially cross-linked cyanopolysiloxane	Rt-2330,SP-2380,SP-2380 (capillary)

TABLE 2.8 (*continued*)

USP Nomenclature Code	USP Stationary-Phase Description	Similar Stationary Phases
G49	Proprietary derivatized phenyl groups on a polysiloxane back bone	A suitable grade is available commercially as Optima Delta 3 from Machery-Nagel,Inc., 215 River Vale Road, River Vale, NJ 07675
G50	Polyethylene glycol,cross-linked (average mol. wt.>20,000)	FAMEWAX"

Source: *USP Column Cross-Reference Charts:* Restek Corporation, Supelco, SGE, and Agilent Technologies.

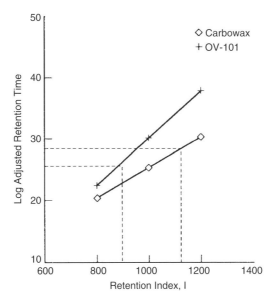

Figure 2.5 Plot of logarithm adjusted retention time versus Kovats' retention index: isoamyl acetate at 120°C

Alternatively, the retention index of an analyte at an isothermal column temperature can be calculated as

$$I = 100N + 100 \left[\frac{\log t'_R(A) - \log t'_R(N)}{\log t'_R(n) - \log t'_R(N)} \right] \tag{2.2}$$

where N and n are the smaller and larger n-paraffin, respectively, that bracket substance A. For temperature-programmed runs, the adjusted retention times in equation 2.2 are replaced by the appropriate elution temperatures in Kelvin. An I value computed by equation 2.2 is strongly recommended because it is inherently more accurate than that obtained by the graphical approach.

The retention index normalizes instrumental variables in gas chromatographs, allowing retention data generated on different systems to be compared and permit interlaboratory comparisons of retention data. For example, isoamy acetate with a retention index of 1128 will elute between n-C_{11} and n-C_{12} under the same chromatographic conditions. The retention index is also very helpful in comparing relative elution orders of a series of analytes on a specific column at a given temperature and in comparing the selective behavior of two or more columns.

McReynolds has tabulated the retention index for a large number of compounds on various liquid phases (59); a review of the retention index system has been prepared by Ettre (60). Postrun calculation of index values is greatly facilitated by using a reporting integrator or data acquisition system. Consistent with the growing trend for computer assistance in gas chromatography is the availability of retention index libraries for drugs and pharmaceuticals; organic volatiles; pesticides, herbicides, and PCBs of environmental significance; methyl esters of fatty acids; food and flavor volatiles; solvents; and chemicals (61).

McReynolds and Rohrschneider Classifications of Stationary Phases

The most widely used system for classifying liquid phases, the McReynolds system (55), has been employed to characterize virtually every stationary phase. McReynolds selected 10 probe solutes of different functionality, each designated to measure a specific interaction with a liquid phase. He analyzed these probe solutes and measured their I values for over 200 phases, including squalane, which served as a reference liquid phase under the same chromatographic conditions. A similar approach had been implemented previously by Rohrschneider (56, 57) with five probes. In Table 2.9 the probes used in both approaches and their functions are

TABLE 2.9 Probes Used in the McReynolds and Rohrschneider Classifications of Liquid Phases

Symbol	McReynolds Probe	Rohrschneider Probe	Measured Interaction
X′	Benzene	Benzene	Electron density for aromatic and olefinic hydrocarbons
Y′	n-Butanol	Ethanol	Proton-donor and proton-acceptor capabilities (alcohols, nitriles)
Z′	2-Pentanone	2-Butanone	Proton-acceptor interaction (ketones, ethers, aldehydes, esters)
U′	Nitropropane	Nitromethane	Dipole interactions
S′	Pyridine	Pyridine	Strong-proton acceptor interaction
H′	2-Methyl-2-pentanol	—	Substituted alcohol interaction similar to n-butanol
J′	Iodobutane	—	Polar alkane interactions
K′	2-Octyne	—	Unsaturated hydrocarbon interaction similar to benzene
L′	1,4-Dioxane	—	Proton-acceptor interaction
M′	cis-Hydrindane	—	Dispersion–interaction

listed. McReynolds calculated a ΔI value for each probe, where

$$\Delta I = I_{\text{liquid phase}} - I_{\text{squalane}}$$

As the difference in the retention index for a probe on a given liquid phase and squalane increases, the degree of specific interaction associated with that probe increases. The cumulative effect, when summed for each of the 10 probes, is a measure of the overall "polarity" of the stationary phase. In a tabulation of McReynolds constants, the first five probes usually appear and are represented by the symbols X′, Y′, Z′ U′, and S′. Each probe is assigned a value of zero, with squalane as the reference liquid phase.

There were several significant consequences resulting from these classification procedures. Phases that have identical chromatographic behavior also have identical constants. In this case the selection of a stationary phase could be based on a consideration such as thermal stability, lower viscosity, cost, or availability. McReynolds constants of the more popular stationary phases for packed columns GC are listed in Table 2.10. Note that DC-200 (a silicone oil of low viscosity) and OV-101 or SE-30 (a dimethylpolysiloxane) have nearly identical constants but also observe that these two polysiloxanes have a more favorable higher-temperature limit. Comparison of this type curtailed the proliferation of phases, eliminated the duplication of phases, and simplified column selection. Many phases quickly became obsolete and were replaced by a phase having identical constants but of higher thermal stability, such as a polysiloxane phase. Today, polysiloxane-type phases are the most commonly used stationary phases for both packed (and capillary column) separations because they exhibit excellent thermal stability, have favorable solute diffusivities, and are available in a wide range of polarities. They are discussed in greater detail in Chapter 3.

There was also an impetus to consolidate the number of stationary phases in use during the mid-1970s. In 1973, Leary et al. (63) reported the application of a statistical nearest-neighbor technique to the 226 stationary phases in the McReynolds study and suggested that just 12 phases could replace the 226. The majority of these 12 phases appear in Table 2.10. Delley and Frederick found that four phases, OV-101, OV-17, OV-225, and Carbowax 20M, could provide satisfactory gas chromatographic analysis of 80% of a wide variety of organic compounds (64). Hawkes et al. (65) reported the findings of a committee effort on this subject and recommended a condensed list of six preferred phases on which almost all gas–liquid chromatographic analyses could be performed: (1) a dimethylpolysiloxane (e.g., OV-101, SE-30, SP-2100), (2) a 50% phenylpolysiloxane (OV-17, SP-2250), (3) polyethylene glycol of molecular weight >4000), (4) DEGS, 5) a cyanopropylpolysiloxane (Silar 10C, SP-2340), and (6) a trifluoropropylpolysiloxane (OV-210, SP-2401). Chemical structures of the more popular polysiloxanes used as stationary phases are illustrated in Figure 2.6.

Another feature of the McReynolds constants is guidance in the selection of a column that will separate compounds with different functional groups, such as ketones from alcohols, ethers from olefins, esters from nitriles, and so on. If an analyst needs a column to elute an ester after an alcohol, the stationary phase

TABLE 2.10 McReynolds Constants Cross-Referenced to Commonly Used Stationary Phases

Phase	Min./Max. Temperature (°C)	Chemical Nature	X'	Y'	Z'	U'	S'	Phases of Similar Structure
Squalane	20/100	Cycloparaffin	0	0	0	0	0	
Polysiloxanes								
DC 200	0/200	Dimethylsilicone	16	57	45	66	43	SP-2100, SE-30, OV-101, OV-1
DC-710	5/250	Phenylmethylsilicone	107	149	153	228	190	OV-11
SE-30	50/300	Dimethyl	15	53	44	64	41	SP-2100, OV-101, OV-1
SE-54	50/300	5% Phenyl, 1% vinyl	33	72	66	99	67	
OV-1	100/350	Dimethyl (gum)	16	55	44	65	42	SP-2100
OV-3	0/350	10% Phenylphenylmethyldimethyl	44	86	81	124	88	
OV-7	0/350	20% Phenylphenylmethyldimethyl	69	113	111	171	128	
OV-11	0/350	35% Phenylphenylmethyldimethyl	102	142	145	219	178	DC-710
OV-17	0/375	50% Phenyl–50% methyl	119	158	162	243	202	SP-2250
OV-22	0/350	65% Phenylphenylmethyldiphenyl	160	188	191	283	253	
OV-25	0/350	75% Phenylphenylmethyldiphenyl	178	204	208	305	280	
OV-61	0/350	33% Phenyldiphenyldimethyl	101	143	142	213	174	
OV-73	0/325	5.5% Phenyldiphenyldimethyl (gum)	40	86	76	114	85	
OV-101	0/350	Dimethyl (fluid)	17	57	45	67	43	SP-2100, SE-30, OV-1
OV-105	0/275	Cyanopropylmethyldimethyl	36	108	93	139	86	
OV-202	0/275	Trifluoropropylmethyl (fluid)	146	238	358	468	310	
OV-210	0/275	Trifluoropropylmethyl (fluid)	146	238	358	468	310	SP-2401

Phase	Temp.	Composition						Equivalent phases
OV-215	0/275	Trifluoropropylmethyl (gum)	149	240	363	478	315	SP-2300, Silar 5 CP
OV-225	0/265	Cyanopropylmethylphenylmethyl	228	369	338	492	386	SP-2340
OV-275	25/275	Dicyanoallyl	629	872	763	1106	849	
OV-330	0/250	Phenyl silicone–Carbowax copolymer	222	391	273	417	368	
OV-351	50/270	Carbowax–nitroterephthalic acid polymer	335	552	382	583	540	SP-1000
OV-1701	0/250	14% Cyanopropylphenyl	67	170	153	228	171	
Silar 5 CP	0/250	50% Cyanopropyl–50% phenyl	319	495	446	637	531	SP-2300, OV-225
Silar 10 CP	0/250	100% Cyanopropyl	520	757	660	942	800	SP-2340
SP-2100	0/350	Methyl	17	57	45	67	43	SE-30, OV-101, OV-1
SP-2250	0/375	50% Phenyl	119	158	162	243	202	OV-17
SP-2300	20/275	50% Cyanopropyl	316	495	446	637	530	OV-225
SP-2310	25/275	55% Cyanopropyl	440	637	605	840	670	
SP-2330	25/275	90% Cyanopropyl	490	725	630	913	778	
SP-2340	25/275	100% Cyanopropyl	520	757	659	942	800	Silar 10 CP
SP-2401	0/275	Trifluoropropyl	146	238	358	468	310	OV-210
Nonsilicone phases								
Apiezon L	50/300	Hydrocarbon grease	32	22	15	32	42	
Carbowax 20M	60/225	Polyethylene glycol	322	536	368	572	510	Superox-4, Superox-20M
DEGS	20/200	Diethylene glycol succinate	496	746	590	837	835	
TCEP	0/175	1,2,3-Tris(2-cyanoethoxy)propane	594	857	759	1031	917	
FFAP	50/250	Free fatty acid phase	340	580	397	602	627	OV-351

Source: Data from refs. 8, 28, and 35.

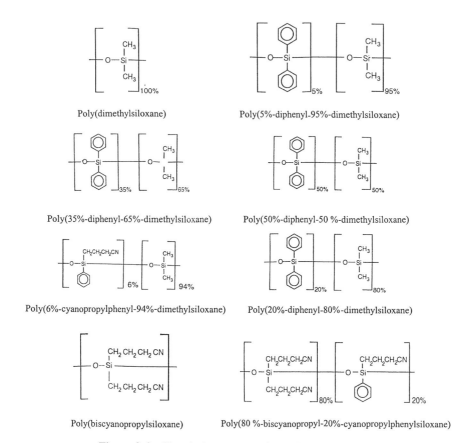

Figure 2.6   Chemical structures of popular polysiloxanes

should exhibit a larger Z′ value with respect to its Y′ value. In the same fashion, a stationary phase should exhibit a larger Y′ value with respect to Z′ if an ether is to elute before an alcohol.

Evaluation of Column Operation

Several parameters can be used to evaluate the operation of a column and to obtain information about a specific system. Using the principles underlying the plate theory discussed by Grob (66), plotting the concentration of solute (in percent) against the volume of mobile phase or the number of plate volumes for the tenth, twentieth, and fiftieth plates in the column, one would obtain a plot such as that shown in Figure 2.7a. Improved separation of component peaks is possible for columns that have a larger plate number. Similar information can be obtained if one plots concentration of solute (percent of total) versus plate number. Figure 2.7b shows the band positions after 50, 100, and 200 equilibrations with the mobile phase.

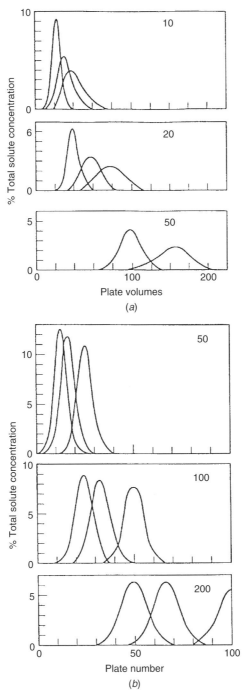

Figure 2.7 (*a*) Elution peaks for three solutes from various plate columns. Top: 10 plates; middle: 20 plates; bottom: 50 plates. (*b*) Plate position of components after variable number of equilibrations. Top: 50 equilibrations; middle: 100 equilibrations; bottom: 200 equilibrations

An ideal gas chromatographic column is considered to have high resolving power, high speed of operation, and high capacity. One of these factors can usually be improved at the expense of another. If we are fortunate, we may in the near future be able to achieve two of the three perhaps all three. Thus, a number of column parameters must be discussed to enable us to arrive at efficient operation of a column. Let us now consider several of these parameters and illustrate them with appropriate relationships.

Column Efficiency.   Two methods are available for expressing the efficiency of a column in terms of HETP (height equivalent to a theoretical plate): measurement of the peak width (1) at the baseline,

$$N = 16 \left(\frac{V_R}{w_b} \right)^2 \tag{2.3}$$

and (2) at half-height,

$$N = 5.54 \left(\frac{V_R}{w_h} \right)^2 \tag{2.4}$$

In determining N, we assume that the detector signal changes linearly with concentration. If it does not, N cannot measure column efficiency precisely. If equation 2.3 or 2.4 is used to evaluate peaks that are not symmetrical, positive deviations of 10 to 20% may result. Since N depends on column operating conditions, these should be stated when efficiency is determined. There are several ways by which one may calculate column efficiency other than by equations 2.3 and 2.4). Figure 2.8 and Table 2.11 illustrate other ways in which this information may be obtained.

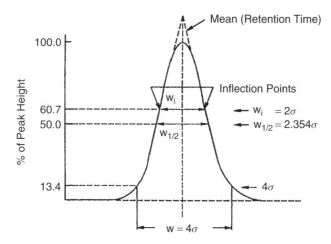

Figure 2.8   Pertinent points on a chromatographic band for the calculation of column efficiency

TABLE 2.11   Calculation of Column Efficiency from Chromatograms

Standard Deviation Terms	Measurements	Plate Number N
$A/h(2\Delta)^{1/2}$	t_R and band area A and height h	$2\Delta(t_Rh/A)^2$
$W_i/2$	t_R and width at inflection points $(0.607\mathrm{h})w_i$	$4(t_R/w_i)^2$
$w_h/(8\ln 2)^{1/2}$	t_R and width at half-height w_h	$5.55(t_R/w_h)^2$
$w_b/4$	t_R and baseline width w_b	$16(t_R/w_h)^2$

Effective Number of Theoretical Plates.   The term *effective number of theoretical plates*, N_{eff}, was introduced to characterize open-tubular columns. In this relationship, adjusted retention volume, V_R', in lieu of total retention volume, V_R, is used to determine the plate number:

$$N_{\text{eff}} = 16\left(\frac{V_R'}{w_b}\right)^2 = 16\left(\frac{t_R'}{w_b}\right)^2 \tag{2.5}$$

The N_{eff} value is useful for comparing a packed and an open-tubular column or two similar columns when both are used for the same separation. Open-tubular columns generally have a larger number of theoretical plates. One can translate regular number of plates, N, to effective number of plates, N_{eff}, by the expression

$$N_{\text{eff}} = N\left(\frac{k}{1+k}\right)^2 \tag{2.6}$$

as well as the ratio of plate height to effective plate height:

$$H_{\text{eff}} = H\left(\frac{1+k}{k}\right)^2 \tag{2.7}$$

Similarly, the number of theoretical plates per unit time can be calculated:

$$\frac{N}{t_R} = \frac{\overline{u}(k)^2}{t_R(l+k)^2} \tag{2.8}$$

where u is the average linear velocity. This relationship accounts for characteristic column parameters, thus offering a way to compare different types of columns.

Resolution.   The separation of two components as the peaks appear on the chromatogram is characterized by

$$R_s = \frac{2\Delta t_R'}{w_{b1} + w_{b2}} \tag{2.9}$$

where $\Delta t'_R = t'_{R2} - t'_{R1}$. If the peak widths are equal (i.e., $w_{b1} = w_{b2}$, equation 2.9 may be rewritten as

$$R_s = \frac{\Delta t'_R}{w_b} \tag{2.10}$$

The two peaks will touch at the baseline when t'_R is equal to 4σ:

$$t'_{R2} - t'_{R1} = \Delta t'_R \tag{2.11}$$

If the two peaks are separated by a distance of 4σ, $R_s = 1.0$. If peaks are separated by 6σ, $R_s = 1.5$.

Resolution may also be expressed in terms of the retention index of two components:

$$R_s = \frac{I_2 - I_1}{w_h f} \tag{2.12}$$

where f is the correction factor (1.699) because $4\sigma = w_b = 1.699 w_h$. A more usable expression for resolution is

$$R_s = \frac{1}{4}(N)^{1/2} \frac{\alpha - 1}{\alpha} \frac{k}{1 + k} \tag{2.13}$$

where N and k refer to the later eluting compound of the pair. Since α and k are constant for a given column (under isothermal conditions), resolution will depend on the number of theoretical plates N. The k term generally increases with a temperature decrease, as does α but to a lesser extent. The result is that at low temperatures, one finds that fewer theoretical plates or a shorter column are required for the same separation.

Required Plate Number.   If the retention factor, k, and the separation factor, α, are known, the required number of plates (n_{ne}) can be calculated for the separation of two components (the k value refers to the more readily sorbed component). Thus,

$$n_{ne} = 16 R_s^2 \left(\frac{\alpha}{\alpha - 1}\right)^2 \left(\frac{1 + k}{k}\right)^2 \tag{2.14}$$

The R_s value is set at the 6σ level, or 1.5. In terms of the required effective number of plates, equation 2.14 would be

$$n_{eff} = 16 R_s^2 \left(\frac{\alpha}{\alpha - 1}\right)^2 \tag{2.15}$$

Taking into account the phase ratio β, we can write equation 2.14 as

$$n_{ne} = 16 R_s^2 \left(\frac{\alpha}{\alpha - 1}\right)^2 \left(\frac{\beta}{k_2 + 1}\right)^2 \tag{2.16}$$

TABLE 2.12   Values for the Last Term of Equation 2.18

k	0.25	0.5	1.0	5.0	10	20	50	100
$\left(\frac{1+k}{k}\right)^2$	25	9	4	1.44	1.21	1.11	1.04	1.02

Equations 2.14 and 2.16 illustrate that the number of plates required will depend on the partition characteristics of the column and the relative volatility of the two components: that is, on K and β. Table 2.12 gives the values of the last term of equation 2.14 for various values of k. These data suggest a few interesting conclusions: If $k > 5$, the plate numbers are controlled mainly by column parameters; if $k > 5$, the plate numbers are controlled by the relative volatility of components. The data also illustrate that k values greater than 20 cause the theoretical number of plates N and effective number of plates N_{eff} to be the same order of magnitude; that is,

$$N \simeq N_{\mathrm{eff}} \qquad (2.17)$$

The relationship in equation 2.14 can also be used to determine the length of column necessary for a separation L_{ne}. We know that $N = L/H$; thus,

$$L_{\mathrm{ne}} = 16R_s^2 H \left(\frac{\alpha}{\alpha - 1}\right)^2 \left(\frac{1+k}{k}\right)^2 \qquad (2.18)$$

Unfortunately, equation 2.18 is of little practical importance because the H value for the more readily sorbed component must be known but is not readily available from independent data.

Let us give some examples of the use of equation 2.14. Table 2.13 gives the number of theoretical plates for various values of α and k, assuming R_s to be at $6\sigma(1.5)$. Using data in Table 2.13 and equation 2.14 we can make an approximate comparison between packed and open-tubular columns. As a first approximation, β values of packed columns are 5 to 30 and for open-tubular columns are 100 to 1000: thus, a ten to hundredfold difference in k. Examination of the data in Table 2.13 shows that when $\alpha = 1.05$ and $k = 5.0$ we would need 22,861

TABLE 2.13   Number of Theoretical Plates for Values of α and k (R_s at $6\sigma = 1.5$)

	α				
k	1.05	1.10	1.50	2.00	3.00
0.1	1, 920, 996	527, 076	39, 204	17, 424	9, 801
0.2	571, 536	156, 816	11, 664	5, 184	2, 916
0.5	142, 884	39, 204	2, 916	1, 296	729
1.0	63, 504	17, 424	1, 296	576	324
2.0	35, 519	9, 801	729	324	182
5.0	22, 861	6, 273	467	207	117
8.0	20, 004	5, 489	408	181	102
10.0	19, 210	5, 271	392	173	97

plates in a packed column, which would correspond to an open-tubular column with $k = 0.5$ having 142,884 plates. Although a greater number is predicted for the open-tubular column, this is relatively easy to attain because longer columns of this type have high permeability and smaller pressure drop than the packed columns.

Separation Factor. The reader will recall that the separation factor, α ($\alpha = K_{Da}/K_{Db} = D_a/D_b = K_a/K_b$, where K_D are distribution constants and D are distribution coefficients) is the same as the relative volatility term used in distillation theory. In 1959, Purnell (67, 68) introduced another separation factor term (S) to describe the efficiency of a column. It can be used very conveniently to describe efficiency of open-tubular columns:

$$S = 16(\frac{V'_R}{w_b})^2 = 16(\frac{t'_R}{w_b})^2 \tag{2.19}$$

where V'_R and t'_R are the adjusted retention volume and adjusted retention time, respectively. Equation 2.19 may be written as a thermodynamic quantity that is characteristic of the separation but independent of the column. In this form we assume resolution, R_s, at the 6σ level or having a value of 1.5. Therefore, from equation 2.15, we obtain

$$S = 36 \left(\frac{\alpha}{\alpha - 1}\right)^2 \tag{2.20}$$

Separation Number. We can also calculate a separation number SN or Trennzahl number TZ as another way of describing column efficiency (69). A separation number represents the number of possible peaks that appear between two n-paraffin peaks with consecutive carbon numbers. It may be calculated by

$$SN = TZ = \frac{t_{R2} - t_{R1}}{w_{h1} + w_{h2}} - 1 \tag{2.21}$$

This equation may be used to characterize capillary columns or the application of programmed pressure or temperature conditions for packed columns. This concept is depicted in Figure 2.9.

Analysis Time. If possible, we like to perform the chromatographic separation in minimum time. Time is important in analysis but is particularly important in process chromatography and in laboratories having a high sample throughput. Analysis time is based on the solute component that is most readily sorbed. Using the equation for determination of retention time, we obtain

$$t = \frac{L(1 + k)}{\overline{u}} = \frac{NH}{\overline{u}}(1 + k) \tag{2.22}$$

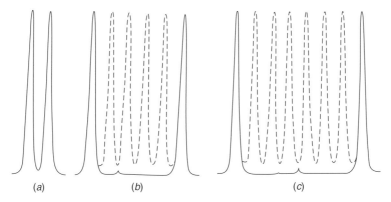

Figure 2.9   Separation number (Trennzahl): (a) Tz 0; (b) Tz 4; (c) Tz 6.

and substituting the value for the required number of plates, N_{ne}, for N [equation 2.14], we arrive at an equation for the minimum analysis time, t_{ne}:

$$t_{ne} = 16R_s^2 \frac{H}{\bar{u}} \left(\frac{\alpha}{\alpha - 1} \right)^2 \frac{(1 + k)^3}{k^2} \tag{2.23}$$

The term $H/\bar{u}$ can be expressed in terms of the modified van Deemter equation:

$$H = A + \frac{B}{u} + Cu \tag{2.24}$$

$$\frac{H}{\bar{u}} = \frac{A}{\bar{u}} + \frac{B}{\bar{u}^2} + C_l + C_g \tag{2.25}$$

For minimum analysis time, high linear gas velocities are used; thus, the first two terms on the right side of equation 2.25 may be neglected. Therefore,

$$\frac{H}{\bar{u}} = C_l + C_g \tag{2.26}$$

Substituting equations 2.14 and 2.26 into equation 2.23, we obtain

$$t_{ne} = N_{ne}(C_l + C_g)(1 + k) \tag{2.27}$$

This equation indicates that minimal separation time depends on plate numbers, capacity factor, and resistance to mass transfer. It should be pointed out that the analysis times calculated from equation 2.23 also depend on the desired resolution. Our example calculations were made on the basis of resolution, $R_s = 1.5$. For a resolution of 1.00, even shorter analysis times can be achieved.

   Figure 2.10 gives a representation of an idealized separation of component zones and the corresponding chromatographic peaks for a three-component system. With columns of increasing number of plates, we see better resolution as column efficiency increases.

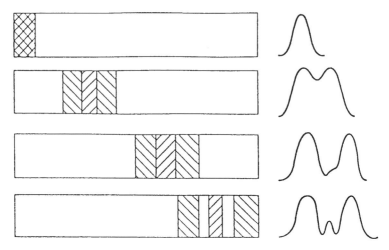

Figure 2.10 Idealized separation process with two major components and one minor component.

Optimization of Packed Column Separations

Examination of the parameters in the van Deemter expression [equation 2.1] term by term provides a basis for optimizing a packed column separation. The plate height, H, of a packed column may be represented as the sum of the eddy diffusion, molecular diffusion, and mass transfer effects. Thus, to attain maximum column efficiency, each term in the plate height equation should be minimized.

Eddy Diffusion Term. In this term, $2\lambda d_p$, λ is a dimensionless packing term, expressing the uniformity with which a packed column is filled and d_p is the diameter of a support particle. Eddy diffusion is also referred to as the *multiple-path effect*; it is minimized by using small particles of support materials. A support of 100/120 mesh produces a more efficient column than 60/80-mesh particles and should be used whenever possible. A support of lower mesh (e.g., 80/100 or 60/80) should be selected to avoid a high pressure drop within a long column. This term is also independent of linear velocity (u) and flow rate, F_c.

Molecular, Longitudinal, or Ordinary Diffusion. In this term, $2\gamma D_g/u$, γ is the tortuosity factor, expressing uniformity of support particle size and shape; D_g is the diffusion coefficient of the solute in the mobile phase; and u is the linear velocity of the mobile phase. The term becomes significant at very low flow rates. This contribution may be minimized by using a carrier gas of high molecular weight (nitrogen, carbon dioxide, or argon) because its diffusion coefficients, D_g, are lower than that of a lower-molecular-weight carrier gas (helium or hydrogen), yielding a lower minimum in an H versus u profile (Figure 2.11). Factors affecting the choice of helium versus hydrogen are discussed in more detail in Chapter 3. However, other factors, such as detector compatibility, can override carrier gas

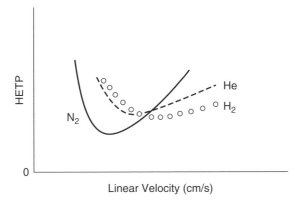

Figure 2.11   Plot of HETP versus linear velocity.

selection; thus, helium is preferred over nitrogen as a carrier gas with a thermal conductivity detector. Colon and Baird (70) have provided excellent coverage of the various detectors used in gas chromatography.

Nonequilibrium or Resistance to the Mass Transfer Contribution.   In this term, $(8/\pi^2)[k(1+k)^2](d_f^2/D_l)u$, π is a constant (3.1416), k the retention or capacity factor, d_f the thickness of the liquid-phase film, D_l the diffusion coefficient in the liquid stationary phase, and u the linear velocity of the mobile phase. The term requires that compromises be made. The magnitude of the term can clearly be minimized by decreasing film thickness (by using packing that has a lower loading of stationary phase). Therefore, column efficiency increases (and time of analysis decreases) with a decrease in stationary-phase loading, as may be seen in Figure 2.4. If a support is too thinly coated, the exposure of active sites on the support may cause adsorption of solutes. A decrease in column temperature lowers the magnitude of the k term; however, lowering of the column temperature also decreases D_l by increasing the viscosity of the stationary phase. The effects of the various changes in chromatographic parameters on resolution that can be implemented are illustrated in Figure 2.12.

A relationship between stationary-phase concentration and column temperature is depicted in Figure 2.13. Decreasing column temperature increases the time of analysis; to have the same analysis time on a heavier-loaded packing in an identical column at the same flow rate requires a higher column temperature.

2.4   COLUMN PREPARATION

Most laboratories today purchase columns of designated dimensions (length, o.d., and i.d.) containing a specified packing (e.g., untreated or treated support coated with a given liquid-phase loading) directly from a chromatography vendor. In-house preparation of packings and filling columns can be time consuming and is

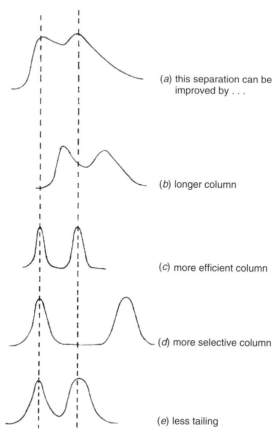

Figure 2.12   Effect of selected column changes on resolution. (From ref. 46, copyright ©
1985 John Wiley & Sons, Inc., with permission.)

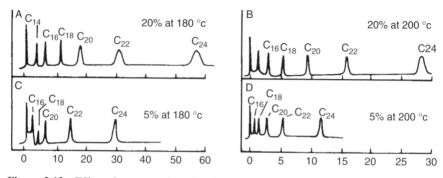

Figure 2.13   Effect of concentration of stationary phase and column temperature on sample
resolution (methyl esters of fatty acids). (From ref. 46, copyright © 1985 John Wiley & Sons,
Inc., with permission.)

false economy; more important, vendors can do the job better. Some supports are difficult to coat uniformly, and some packings present a problem when filling a column. Nevertheless, presented below are guidelines and concise descriptions of procedures recommended for preparing a packed column.

Coating Methods

The techniques of solvent evaporation and solution coating are the most commonly used procedures for deposition of a liquid phase on a support. Solvent evaporation is employed for coating supports with high concentrations (>15%) of viscous phases; the solution-coating method produces a more uniform phase deposition and is more widely utilized.

In solvent evaporation, a known amount (by weight) of stationary phase is dissolved in an appropriate solvent. A weighed amount of support is added to the solution and the solvent is allowed to evaporate slowly from the slurry. Since all stationary phase is deposited on the support, stirring or thorough mixing is a necessity; otherwise, a nonuniform deposition of phase will result.

The technique of solution-coating consists of the following steps:

1. A solution of known concentration of liquid phase in its recommended solvent is prepared.
2. To a known volume of this solution is added a weighed amount of solid support.
3. The resulting slurry is transferred into a Büchner funnel, with the remaining solvent removed by vacuum.
4. The volume of filtrate is measured.
5. After suction is completed, allow the "wet" packing to air-dry on a tray in a hood to remove residual solvent. Do *not* place damp packing into a laboratory drying oven!
6. The mass of liquid phase retained on the support is computed since the concentration of liquid phase in the solution is known.

This technique produces a uniform coating of a support, minimum generation of fines from the support particles, and minimum oxidation of the stationary phase. Further details regarding these procedures are described in reference 46.

Tubing Materials and Dimensions

Column tubings have been made of copper (no longer used), aluminum (use has diminished over the years), stainless steel, stainless steel lined with fused silica, and glass. The nature and reactivity of the sample will govern the choice of tubing for packed column GC. Of the materials available, glass is the most inert and the best material for most applications, although somewhat fragile. Glass columns are used where the sample components may have interactions with metal column walls.

If your analyses involve reactive compounds, you could use SilcoSmooth tubing (Restek), which is made from ultrasmooth seamless 304 stainless steel and treated with Restek's innovative Silcosteel deactivation process. SilcoSmooth tubing can replace glass columns for virtually any application. Analyzing ppb levels of sulfur compounds has been virtually impossible using stainless steel columns. Restek's Sulfinert tubing combined with TX-XLSulfur packing material is ideal for trace-level sulfur compound analyses. Both SilcoSmooth and Sulfinert columns offer the inertness of glass without the breakage problems. Overtightening a fitting attached to a glass column can cause the dreaded "ping" sound of broken glass. Utilization of a special torque wrench, which breaks apart itself rather than the glass column when a specific torque level is exceeded, is a good investment for a gas chromatographic laboratory. Glass columns rarely cause tailing or decomposition of sample components. Empty glass columns need deactivation or silylation prior to packing. Usually, this is accomplished by filling a *thoroughly* cleaned empty glass column with a 5% solution of dimethyldichlorosilane in toluene. After standing for 30 minutes, the column is rinsed successively with toluene and methanol, then purged with dry nitrogen, after which it is ready to be packed. Moreover, as opposed to a metal column, glass permits direct visualization of how well a column is packed after filling and also after the column has been used for separations. Teflon tubing, also inert, is used for the analysis of sulfur gases, halogens, HF, and HCl, but it has a temperature limitation of $250°C$ (softening may begin to occur at $225°C$, and Teflon tubing is also permeable to gases; it is used for making permeation tubes).

Nickel tubing offers the attractive combination of the durability and strength of metal tubing with the favorable chemical inertness of glass. It is often used for the analysis of caustic or oxidizing compounds or gases. Hastelloy tubing (Restek) is a nickel–chromium alloy with excellent inertness. It is usually employed only for highly corrosive or oxidizing compounds or gases. Stainless steel is the next-least reactive material and is utilized for analysis of hydrocarbons, permanent gases, and solvents. As is the case with all metal tubing, stainless steel columns should be rinsed with nonpolar and polar solvents to remove residual oil and greases. When used for the analysis of polar species, a higher grade of stainless steel tubing with a polished inner surface is recommended. Copper and aluminum tubing have been employed for noncritical separations, but their use is not recommended and should be restricted to plumbing of cylinder-instrument gas lines. Oxide formation can occur on the inner surface of these materials (copper and aluminum), resulting in adsorptive tailing and/or catalytic problems under chromatographic conditions.

Glass Wool Plugs and Column Fittings

Chromatographic packings are retained within a column by a wad or plug of glass wool or fritted metal plugs. Since the chemical nature of the wool closely resembles that of the glass column, it should be deactivated by the same procedure used as that for glass columns. It is advisable to soak the wool further in a dilute solution of phosphoric acid (H_3PO_4) for the analysis of acidic analytes such as phenols and fatty acids. Untreated or improperly treated glass wool exhibits an

TABLE 2.14   Ferrule Materials for Packed Columns

Material	Temperature Limit ($^\circ$C)	Properties
Metal		
Brass	250	Permanent connection on metal columns.
Stainless steel	450	Permanent connection on metal columns.
Teflon	250	Low upper temperature limit and cold-flow properties render this material unsuitable for temperature programming and elevated temperature operation; reusable to some extent.
Ceramic-filled	250	Isothermal use only; conforms easily to glass. Used for connections to mass spectrometers.
Graphite (G)	450	High-temperature limit with no bleed or decomposition; soft and easily deformed upon compression; may be resealed only a limited number of times.
Vespel 100% polyimide (PI)	350	Good reusability factor; can be used with glass, metal, and Teflon columns; may seize on metal and glass columns when used at elevated temperatures.
Vespel/graphite 85% PI, 15% G	400	Excellent reusability; will not seize to glass or metal; performs better than graphite and Vespel alone; 60% PI composite seals with less torque and has added lubricity.

active surface and can cause peak tailing. Alternatives to glass wool are stainless steel frits and screens for gas chromatographic purposes, available from vendors of chromatographic supplies. These should be treated with polar and nonpolar solvents to remove residual oil and grease.

Ferrules and metal retaining nuts are used to form leaktight seals of a column in a gas chromatograph. Criteria for selection of the proper ferrule material are column diameter, column tubing material, maximum column temperature, and whether the connection is designated for a single use or for multiple connections and disconnections. Ferrules fabricated from various materials for metal-to-metal, glass-to-metal, and glass-to-glass connections are commercially available for use with $\frac{1}{16}-$, $\frac{1}{8}-$, and $\frac{1}{4}$-in.-o.d. packed columns. The properties and characteristics of common types are presented in Table 2.14.

Filling the Column

With practice, the following procedure for packing columns can produce the desired goal of a tight packing bed with minimum particle fracturing. First, a metal column is precoiled for easy attachment to the injector and detector of the instrument in which it is to be installed, or a coiled glass column configured for a specific instrument is procured from a vendor. Figure 2.14 illustrates several representative glass column configurations. Insert a large wad of glass wool partially into one end

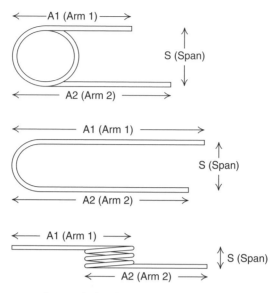

Figure 2.14 Representative packed column configurations. (Courtesy of Alltech Associates.)

of the column, align the excess wool along the outside of the tubing, overlap the excess wool with vacuum tubing, and attach the other end of the vacuum tubing to a faucet aspirator or pump (4). After securing a small funnel to the other end of the column, add packing material to the funnel in small incremental amounts, and gently tap the packing bed while suction is applied. After the column has been packed completely, insert a small piece of silanized glass wool into the inlet end of the column, shut down the vacuum source, remove the vacuum tubing and the large wad of glass wool, and replace the glass wool with a smaller plug of silanized glass wool. This approach eliminates the exasperating sight of packing material zipping out of the column during filling if a sufficiently tight wad of silanized glass wool was not inserted initially into the outlet end of the column.

Conditioning the Column and Column Care

Before a column is used for any analysis, it must be conditioned thermally by heating the column overnight at an oven temperature below the upper limit of the stationary phase with a normal flow rate of carrier gas. The column should *not* be connected to the detector during the conditioning process. The purpose of conditioning is the removal from the column of any remaining residual volatiles and low-boiling species present in the stationary phase which could produce an unsteady baseline at elevated column temperatures, commonly referred to as *column bleed*, and contaminate the detector.

Conditioning a column also helps in redistribution of the liquid phase on the solid support. The degree of conditioning depends on the nature and amount of liquid phase in the column; usually, heating a column overnight at an appropriate elevated temperature produces a steady baseline under chromatographic conditions the following day. Analyses using the more sensitive detectors (ECD, NPD, MS) may require an even longer column conditioning period. Appendix A provides representative chromatograms for many packed column applications. These chromatograms, which appeared originally in *The Packed Column Application Guide*, list the packing, column dimensions, temperature, mobile phase and its flow rate, detector employed and temperature, and the sample size (71).

The following guidelines can prolong the lifetime of a column:

1. Any gas chromatographic column, new or conditioned, packed or capillary, should be purged with dry carrier gas for 15 to 30 minutes before heating to a final elevated temperature to remove the detrimental presence of air.

2. A column should not be heated to an elevated temperature rapidly or ballistically but should be heated by slow to moderate temperature programming to the final temperature desired.

3. Excessively high conditioning and operating temperatures reduce the lifetime of any gas chromatographic column,

4. Use "dry" carrier gas or install a moisture trap in the carrier gas line. Do not inject aqueous samples on a column containing a stationary phase that is intolerant of water.

5. The accumulation of high-boiling compounds from repetitive sample injections occurs at the inlet end of the column and results in discoloration of the packing. It is a simple matter to remove the discolored segment of packing and replace it with fresh packing material. This action prolongs the column lifetime.

6. Do not thermally shock a column by disconnecting a column while hot. Allow the column to cool to ambient temperature prior to disconnection. Packings are susceptible to oxidation when hot.

7. Cap the ends of a column for storage to prevent air and dust particles from entering the column. Save the box in which a glass column was shipped for safe storage of the column.

2.5 *UNITED STATES PHARMACOPEIA* AND *NATIONAL FORMULARY* CHROMATOGRAPHIC METHODS

Table 2.15 lists representative methods recommended for various pharmaceuticals and residual solvents in pharmaceuticals. Refer to Tables 2.6 and 2.8 for an explanation of the supports and liquid phases shown in Table 2.15. These methods are reviewed and updated continually. The methods listed in Table 2.15 are current as of the first quarter of 2006.

TABLE 2.15 Gas Chromatographic Methods Described in USP/NF Monographs[a]

Name of Monograph	Analyte	Packed (P) or Capillary (C)	Column i.d. (mm)	Column Length (m)	Liquid Phase (%)[b]	Support	Column Temperature; Detector; Carrier Gas
Alcohol determination < 611 >	Ethanol	P	4	1.8	NA	S3	120°C; FID; He/N$_2$
Articles of botanical origin—test for pesticides < 561 >	Quantitative analysis of organophosphorus insecticides	C	0.32	30	0.25-μm G1	Fused silica	80–280°C; AFID/FPD; H$_2$, He/N$_2$
	Quantitative analysis of organochlorine and pyrethroid insecticides	C	0.32	30	0.25-μm G1	Fused silica	80–280°C; ECD; H$_2$, He/N$_2$
Barbiturate assay < 361 >	Barbiturate or barbituric acid assay as per monograph	P	4	0.9	3% G10	S1A	200°C; FID; N$_2$
Fats and fixed oils < 401 > (see note for fatty acid mixture)	Fatty acid composition C8–C22 with 1–3 double bonds	C	0.53	30	1.0-μm G16	Fused silica	70–240°C; FID; He
Organic volatile impurities < 467 >	Use of headspace apparatus that transfers a measured amount of headspace automatically is allowed						
	Water-soluble articles	C	0.32 or 0.53	30	1.8- or 3.0-μm G45	Fused silica	30–240°C; FID; He/N$_2$
	Procedure A Procedure B	C	0.32 or 0.53	30	1.8-μm G44	Fused silica	50–165°C; FID; He/N$_2$
	Procedure C	C	0.32 or 0.53	30	1.8- or 3.0-μm G45	Fused silica	40–240°C; FID; He/N$_2$

	Water-insoluble articles	C	0.32 or 0.53	30	1.8- or 3.0-μm G45	Fused silica	30–240°C; FID; He/N$_2$
	Procedure A Procedure B	C	0.32 or 0.53	30	1.8- or 3.0-μm G45	Fused silica	30–240°C; FID; He/N$_2$
	Procedure C	C	0.32 or 0.53	30	1.8- or 3.0-μm G45	Fused silica	30–240°C; FID; He/N$_2$
Method I		C	0.53	30	5-μm G27	Fused silica	35–260°C; FID; He/N$_2$
Method IV	Same as Method V except use heated syringe						
Method V		C	0.53	30	3-μm G43	Fused silica	40–240°C; FID; He
Method VI	Column and column temperature.	P					FID; He/N$_2$
Acetyltributyl Citrate, NF	Assay	C	0.32	30	0.5-μm G42	Fused silica	80–220°C; FID; He
Alcohol USP	Volatile impurities	C	0.32	30	1.8-μm G43	Fused silica	40–240°C; FID; He
Alprazolam USP	Assay	P	3	120 cm	3% G6	S1AB	240°C; FID; He
Amantadine Hydrochloride Capsules, USP	Dissolution content uniformity, assay	P	2	1.22	10% G1	S1A	115°C; FID; He
Amantadine Hydrochloride, USP	Purity	C	0.53	30	1.0-μm G27	Fused silica	70–250°C; FID; He
Amiloxate, USP	Assay	C	0.32	25	0.1-μm G1	Fused silica	60–240°C; FID; He

(Continued)

61

TABLE 2.15 (*continued*)

Name of Monograph	Analyte	Packed (P) or Capillary (C)	Column i.d. (mm)	Column Length (m)	Liquid Phase (%)[b]	Support	Column Temperature; Detector; Carrier Gas
Amitraz. USP	Identification, assay	P	4	1.5	3% G1	S1A	250°C; FID; N$_2$
Ampicillin Sodium, USP	Limit of methylene chloride	P	4	1.8	10% G39	S1A	65°C; FID; N$_2$
Amyl Nitrite, USP	Content of total nitrites	P	3	2	25% methyl-polysiloxane	Calcined diatomite	80°C; TCD; He
	Organic volatile impurities, Method V < 467 >						
Amylene Hydrate, USP	Assay	P	4	2		S2	190°C; TCD; He
Anhydrous Lactose, NF	Content of α and β anomers	P	2	0.9	3% G19	S1A	215°C; FID; He
Antimicrobial agents—content < 341 >	Benzyl alcohol	P	3	1.8	5% G16	S1A	140°C; FID; He
	Chlorobutanol	P	2	1.8	5% G16	S1A	110°C; FID; He/N$_2$
	Phenol	P	3	1.2	5% G16	S1A	145°C; FID; He/N$_3$
	Parabens	P	2	1.8	5% G2	S1A	150°C; FID; He/N$_4$
Atovaquone, USP	Limit of residual organic solvents	P	4	2.8	10% G16	S2	180°C; FID; N$_2$
Atropine Sulfate Ophthalmic Solution, USP	Assay	P	2	1.8	3% G3	S1AB	225°C; FID; N$_2$
Azobenzene, USP	Assay chromatographic purity	C	0.32	25	G1	Fused silica	200–280°C; FID; He
Baclofen, USP	Organic volatile impurities, Method IV < 467 >						

Name	Test	Type			Packing	Support	Conditions
Belladonna Extract, USP	Assay scopolamine and atropine	P	4	1.2	3%G3	S1AB	215°C; FID; He
	Organic volatile impurities, Method IV <467>						
Benzocaine and Menthol Topical Aerosol, USP	Assay for menthol	P	2	1.8	10% G16	S1AB	170°C; FID; He
Benzyl Alcohol, NF	Related compounds	C	0.32	30	0.5-μm G16	Fused silica	50–220°C; FID; He
Brompheniramine Maleate, USP	Related compounds	P	4	1.2	3% G3	S1AB	190°C; FID; He
	Organic volatile impurities, Method I <467>						
Bupivacaine Hydrochloride, USP	Limit of residual solvents	P	4	2	NA	S3	175°C; FID; N$_2$
Butabarbital, USP	Assay	P	4	1.8	10% G37	S1AB	260°C; FID; N$_2$
	Organic volatile impurities, Method V <467>						
Butane, NF	Assay	P	3	6	10% G30S1C	NAWS1	33°C; TCD; He
Butorphanol Tartrate, USP	Assay	P	4	1.8	3% G3	S1AB	250°C; FID; N$_2$
Butyl Alcohol, NF	Impurity butyl ether	P	6	2	25% G29	S1C	85°C; TCD; He
Butylated Hydroxyanisole, NF	Assay	P	2	1.8	10% G26	S1A	180°C; FID; He
	Organic volatile impurities, Method V <467>						
Camphorated Phenol Topical Gel, USP	Assay	P	2	1.8 glass	15% G44	S1A	140°C; FID; He

(Continued)

TABLE 2.15 (*continued*)

Name of Monograph	Analyte	Packed (P) or Capillary (C)	Column i.d. (mm)	Column Length (m)	Liquid Phase (%)[b]	Support	Column Temperature; Detector; Carrier Gas
Camphorated Phenol Topical Solution, USP	Assay, container headspace content	P	2	1.8	15% G44	S1A	140°C; FID; He
Caprylocaproyl macrogolglycerides; also called Caprylocaproyl Polyoxylglycerides NF	Limit of free ethylene oxide and dioxane	C	0.32	30	1.0-μm G1	Fused silica	180–230°C; FID; He
Carbamazepine Extended-Release Tablets	Limit of residual solvents	P	2	3	0.2% G39	S7	75–155°C; FID; He
Carbomer 934P NF	Limit of benzene	C	0.53	30	3.0-μm G43	Fused silica	40–240°C; FID; He
Carbomer Copolymer, NF	Organic volatile impurities, method IV < 467 > Limit of ethyl acetate and cyclohexane	P	2	3	1% G25	S12	115–175°C; FID; He
Carbon Monoxide C11, USP		P		As needed to separate CO_2 from N_2 and CO		Molecular Sieve S3	TCD; radioactivity detector
Castor Oil Emulsion, USP		P	4	1.8	4% G25	S1	245°C; FID; He
Cefoxitin Sodium, USP	Limit of acetone and methanol	P	6.3	1.8	NA	S2	110°C; FID; N_2

Cellaburate, NF	Acetyl and butyryl content	C	0.53	30	1-μm G35	Fused silica	125°C; FID; He
Cetostearyl Alcohol, NF	Assay	P	3	2	10% G2	S1A	205°C; FID; He
Cetyl Alcohol, NF	Assay	P	3	2	10% G3	S1A	205°C; FID; He
Chamomile, Dietary Supplement	Content of bisabolan derivatives	C	0.32	30	0.25-μm G16	Fused silica	70–230°C; FID; He
Chlorhexidine Gluconate Oral Rinse, USP	Content of alcohol	C	0.53	30	1.5-μm G27	Fused silica	35–225°C; FID; He
Chloroxylenol, USP	Related compounds	P	4	1.8	3% G16	S1A	180°C; FID; N_2
Chlorpheniramine Maleate, USP	Related compounds	P	4	1.2	3% G3	S1AB	190°C; FID; He
Choline Chloride DS	Organic volatile impurities, Method IV < 467 >						
	Organic volatile impurities, Method IV < 467 >						
Ciclopirox Olamine Cream, USP	Content of benzyl alcohol (if present)	P	4	2	3% G3	S1AB	100°C; FID; N_2
Cilastatin Sodium, USP	Limit of solvents	C	0.53	30	1-μm G16	Fused silica	50–70°C; FID; He
Clavulanate Potassium, USP	Limit of aliphatic amines	C	0.53	50	5-μm G41	Fused silica	35–150°C; FID; He
	Limit of 2-ethylhexanoic acid	C	0.53	25	1-μm G35	Fused silica	40–200°C; FID; H_2
Clindamycin Palmitate Hydrochloride, USP	Assay	P	3	0.6	1% G36	S1AB	290°C; FID; He
Clioquinol and Hydrocortisone Cream, USP	Assay	P	2	1.8	3% G3	S1AB	165°C; FID; He

(Continued)

TABLE 2.15 (*continued*)

Name of Monograph	Analyte	Packed (P) or Capillary (C)	Column i.d. (mm)	Column Length (m)	Liquid Phase (%)b	Support	Column Temperature; Detector; Carrier Gas
Clofibrate, USP	Purity Limit of p-chlorophenol Organic volatile impurities, Method V < 467 >	C	0.53	15	1.5-μm G1	Fused silica	120–180°C; FID; He
Cocoa Butter, NF	Fatty acid composition Organic volatile impurities, Method IV < 467 >	C	0.25	15	0.25-μm G19	Fused silica	180–240°C; FID; He butter
Cod Liver Oil, USP	Fatty acid profile	C	0.25	30	0.25-μm G16	Fused silica	170–225°C; FID; He
Colchicine, USP	Limit of ethyl acetate Organic volatile impurities, Method I < 467 >	P	4	1.5	20% G14	S1	75°C; FID; N_2
Colestipol Hydrochloride, USP	Identification Organic volatile impurities, Method IV < 467 >	P	3	180 cm	0.25% KOH, 5% G16	S1A	85°C; FID; He
Compound Undecylenic Acid Ointment, USP	Assay	P	2	1.8, glass	3% G1	S1A	165°C; FID; He
Conjugated Estrogens, USP	Identification assay, content of 17α-dihydroequilin, 17β-dihydroequilin, and 17 alpha-estradiol (concomitant components)—	C	0.25	15	Fused silica	0.25-μm G19	220°C; FID; H_2

Limits of
17α-dihydroequilenin,
17β-dihydroequilenin, and
equilenin (signal
impurities)—

Limits of 17β-estradiol and
Δ8,9-dehydroestrone

Limit of estrone, equilin, and
17α-dihydroequilin (free
steroids)

Organic volatile impurities,
Method V 467

Monograph	Test						Conditions
Cyclomethicone, USP	Assay	P	3	3.66	20% G1	S1A	125–320°C; TCD; He
Cyclosporine Injection–Ethanol, USP		P	2	2	NA	S3	145–270°C; FID; N$_2$
Cyclosporine Injection, USP	Alcohol content (where present)	P	2	2	NA	S3	145–270°C; FID; N$_2$
Dehydrated Alcohol	Volatile impurities	C	0.32	30	1.8-μm G44	Fused silica	40–240°C; FID; He
Desflurane USP	Assay	P	2.4	3.7, SS coated polytef.	10% G31, 15% G18	S1A	80–175°C; FID; He
	Related compounds	P	2.4	6.1	25% G16	S1A	75°C; FID; He
Desoximetasone Gel, USP	Alcohol determination, Method II < 611 >	P	4	1.8	NA	S3	120°C; FID; He/N$_2$
Dexamethasone Sodium Phosphate, USP	Alcohol content, Method II < 611 >	P	4	1.8	NA	S3	120°C; FID; He/N$_2$

(Continued)

TABLE 2.15 (*continued*)

Name of Monograph	Analyte	Packed (P) or Capillary (C)	Column i.d. (mm)	Column Length (m)	Liquid Phase (%)[b]	Support	Column Temperature; Detector; Carrier Gas
Dexamethasone Sodium Phosphate, USP	Organic volatile impurities, Method IV < 467 >						
Dexbrompheniramine Maleate, USP	Related compounds	P	4	1.2	3% G3	S1AB	190°C; FID; He
	Organic volatile impurities, Method I < 467 >						
Dexchlorpheniramine Maleate Tablets, USP	Dissolution	P	2	1.8	1.2% G16, 0.5% KOH	S1AB	205°C; FID; He
Dexpanthenol Preparation, USP	Silated pantolactone	P	2.0	1.8	5% G3	S1A	170°C; FID; He/N$_3$
Dextran 40, USP	Limit of alcohol and related impurities	P	2.0	1.8	NA	S3	160°C; FID; N$_2$
Dibutyl Phthalate, NF	Related impurities	P	4	1.5	3% G3	S1A	190°C; FID; N$_2$
Dibutyl Sebacate, NF	Assay	P	2	1.8	10% G41	S1A	150–280°C; FID; He
Dichlorodifluoromethane, NF	Assay and purity	P	2	1.8	1% G25	S12	70–170°C; FID; He
Diethylene Glycol Monoethyl Ether, NF	Limit of free ethylene oxide	C	0.32	30	1.0-μm G1	Glass/quartz	50–230°C; FID; He
Diethylene Glycol Monoethyl Ether, NF	Assay and limit of 2-methoxyethanol, 2-ethoxyethanol, ethylene glycol, and diethylene glycol	C	0.32	30	1.0-μm G46	Fused silica	120–225°C; FID; He

Dihydrotachysterol, USP	Organic volatile impurities, Method IV < 467 >						
Dihydroxyaluminum Sodium Carbonate, USP	Isopropyl alcohol	P	3	0.9	NA	S3	180°C; FID; He/N$_2$
	Organic volatile impurities, Method IV 467						
Dimethyl Sulfoxide Gel, USP	Assay	P	4	1.8	15% G39	S1A	160°C; FID; He
Dimethyl Sulfoxide Irrigation, USP	Proceed as directed in the test for Dimethyl sulfone under *Dimethyl sulfoxide*						
Dimethyl Sulfoxide Topical Solution, USP	Assay	P	4	1.8	10% G16	S1A	170°C; FID; He
Dimethyl Sulfoxide, USP	Limit of dimethyl sulfone	P	3	1.5	10% G25	S1A	100–170°C; FID; He
Dimethylaniline < 223 >	Assay	P	2	2	3% G3	S1A	120°C; FID; N$_2$
Diphenoxylate Hydrochloride and Atropine Sulfate Oral Solution, USP	Assay for atropine sulfate	P	4	1.2	3% G2	S1	230°C; FID; He
	Alcohol content < 611 >						
Doxorubicin Hydrochloride, USP	Limit of solvent residues (as acetone and alcohol)	P	4	2	8–10% G16, 2% KOH	S1A	60°C; FID; He
Doxylamine Succinate, USP–Volatile Related Compounds		P	4	2	5%G16	S1A	212°C; FID; He

(Continued)

TABLE 2.15 (continued)

Name of Monograph	Analyte	Packed (P) or Capillary (C)	Column i.d. (mm)	Column Length (m)	Liquid Phase (%)[b]	Support	Column Temperature; Detector; Carrier Gas
Doxylamine Succinate, USP	Volatile related compounds	P	4	2	5% G16, 5% G12	S1A	212°C; FID; He
	Organic volatile impurities, Method I < 467 >						
Dyphylline Elixir/Dyphylline Oral Solution, USP	Alcohol content	P	4	75 cm	20% G20	S1AB	85°C; FID; N_2
Echothiophate Iodide for Ophthalmic Solution, USP	Water	P	2	1.8	NA	S3	115°C; TCD; He
Enflurane, USP	Assay	P	4	3, SS	20% G4	S1A	60–125°C; TCD; He
Enzacamene, USP	Assay, Related compounds	C	0.32	30	0.25-μm G1	Fused silica	100–300°C; FID; He
Ethchlorvynol Capsules, USP	Assay	P	4	1.8	10% G16	S1AB	160°C; FID; He
Ethchlorvynol, USP	Assay and purity	P	4	1.8	10% G16	S1AB	160°C; TCD; N_2
Ether, USP	Low-boiling hydrocarbons	P	2	3.7, SS	30% G22	S1C	80°C; FID; N_2
Ethyl Acetate, NF	Chromatographic purity	P	4	1.8	NA	S11	115–200°C; FID; He/N_2
	Organic volatile impurities, Method I < 467 > modified	C	0.53	30	1-μm G16	Fused silica	50–230°C; FID; He/N_2
Ethylcellulose Aqueous Dispersion, NF	Identity	P	2	1.8	10% G1	S1A	220°C; FID; H_2/N_2

(Continued)

Article	Determination				Packing	Support	Conditions
Ethylcellulose, NF	Organic volatile impurities, Method V <467>						
Etodolac, USP	Assay	P	2	5, SS	3% G2	S1A	80°C; FID; N_2
	Limit of alcohol and methanol	C	0.32	25	5-μm G36	Fused silica	45–280°C; FID; He
Eucalyptol, USP	Assay	C	0.32	60	G16	Fused silica	60–200°C; FID; He
Fenoldopam Mesylate, USP	Limit of residual solvents	C	0.32	30	1.8-μm G43	Fused silica	40–180°C; FID; He
Fludarabine Phosphate, USP	Limit of ethanol	C	0.25	30	1.4-μm G43	Fused silica	40–220°C; FID; He
Fludeoxyglucose F18 Injection, USP	Residual solvents	C	0.53	30	0.25-μm X-linked G16	Fused silica	40–130°C; FID; He
Fluocinonide Topical Solution, USP	Alcohol content	P	2	1.8	NA	S3	130°C; FID; N_2/H_2
Fluticasone Propionate, USP	Bromofluoromethane content	C	0.32	25	5-μm G27	Fused silica	40–200°C; ECD; N_2
	Acetone content	C	0.53	25	2-μm G15	Fused silica	60–180°C; FID; He/N_2
Gadodiamide, USP	Limit of acetone, ethyl alcohol, and isopropyl alcohol	C	0.32	30	1.8-μm G43	Fused silica	40°C; FID; He
Gadoversetamide, USP	Limit of residual solvents	C	0.53	30	1-μm G35	Fused silica	35–110°C; FID; He
Gentamicin Sulfate, USP	Limit of methanol	P	4	1.5	NA	S3	130°C; FID; N_2
Glycerin, USP	Limit of diethylene glycol and related compounds	C	0.53	30	3-μm G43	Fused silica	100–220C; FID; He
	Organic volatile impurities, Method IV <467>						

TABLE 2.15 (continued)

Name of Monograph	Analyte	Packed (P) or Capillary (C)	Column i.d. (mm)	Column Length (m)	Liquid Phase (%)[b]	Support	Column Temperature; Detector; Carrier Gas
Glyceryl Behenate, NF	Identification	P	4	1.8	10% G7	S1A	225°C; FID; He/N$_2$
	Organic volatile impurities, Method IV <467>						
Glyceryl Monostearate, USP	Free glycerine		4	2.4	2% G16	S1A	195°C; FID; He
	Organic volatile impurities, Method IV <467>						
Gold Sodium Thiomalate, USP	Limit of alcohol	C	0.53	30	3-μm G43	Fused silica	40–240°C; FID; He
Gonadorelin Hydrochloride, USP	Water	P	2	180	NA	S3	100°C; TCD; He
Guaifenesin and Codeine Phosphate Oral Solution, USP	Alcohol content (if present)	P	2	1.8	5% G16	S1A	50–100°C; FID; He
Guaifenesin and Codeine Phosphate Oral Solution, USP	Assay for codeine phosphate	P	2	0.6	3% G3	S1A	210°C; FID; He
Guaifenesin and Codeine Phosphate Syrup, USP	Assay for guaifenesin	P	4	1.2	3% G6	S1A	170°C; FID; He
	Alcohol content (if present)	P	2	1.8	5% G16	S1A	50–100°C; FID; He
Guanabenz Acetate, USP	Assay for codeine phosphate	P	2	0.6	3% G3	S1A	210°C; FID; He
	Assay for guaifenesin	P	4	1.2	3% G6	S1A	170°C; FID; He
	Limit of 2,6-dichlorobenzaldehyde	P	3	1.8	20% G1	S1A	190°C; FID; N$_2$

Hexachlorophene USP	Organic volatile impurities, Method V <467> Limit of 2,3,7,8-tetrachlorodibenzo-p-dioxin	P	2	1	G1	S1	250°C; MS; He
Homosalate, USP	Assay	C	0.53	30	1-µm G27	Fused silica	70–220°C; FID; H_2
Hydrocortisone and Acetic Acid Otic Solution, USP	Assay for acetic acid	P	2	1.8	20% G35	S1A	115–190°C; FID; N_2
Hyoscyamine Tablets, USP	Assay	P	2	1.8	3% G3	S1AB	225°C; FID; N_2
Hypromellose, USP	Assay Organic volatile impurities, Method IV <467>	P	4	1.8	20% G28	S1C	130°C; TCD; He
Ifosfamide, USP	Limit of 2-chloroethylamine hydrochloride	P	2	1.8	10% G16, 2% KOH	S1A	140°C; FID; N_2
Imipenem, USP	Solvents	P	3	1.8	10% G16	S5	70–170°C; FID; He
Indinavir Sulfate, USP	Content of alcohol	C	0.53	30	1-µm G14	Fused silica	35–140°C; FID; He
Indomethacin Sodium, USP	Limit of acetone	P	3	1.8	NA	S3	165°C; FID; N_2
Iodixanol, USP	Limit of methanol, isopropyl alcohol, and methoxyethanol	C	0.54	30	1-µm G16	Fused silica	40–100°C; FID; He
Iohexol, USP	Limit of methanol, isopropyl alcohol, and methoxyethanol	C	0.53	30	3-µm G43	Fused- silica	40–100°C; FID; He

(Continued)

73

TABLE 2.15 (*continued*)

Name of Monograph	Analyte	Packed (P) or Capillary (C)	Column i.d. (mm)	Column Length (m)	Liquid Phase (%)[b]	Support	Column Temperature; Detector; Carrier Gas
Iohexol, USP	Limit of 3-chloro-1,2-propanediol	C	0.32	30	1-μm G46	Fused silica	50–200°C; FID; He
Iopromide, USP	Limit of alcohol	C	0.25	30	1.4-μm G43	Fused silica	40–220°C; FID; He
Ioxilan, USP	Organic volatile impurities, Method IV <467> Residual methanol	C	0.53	30	Coated with S2	Fused silica	45–80°C; FID; He
Isoflurane, USP	Assay related compounds	P	2.4	3.7, Ni or SS	10% G31, 15% G18	S1C NaOH, washed	65–110°C; FID; He
Isoflurophate, USP	Assay	P	4	1.8	5% G33	S1AB	80°C; FID; He
Isopropyl Myristate, NF	Assay	P	4	1.8	10% G8	S1A	90–210°C; FID; N$_2$
Isopropyl Palmitate, NF	Assay	P	4	1.8	10% G8	S1A	90–210°C; FID; N$_2$
Isosorbide Concentrate, USP	Organic volatile impurities, Method IV <467> Methyl ethyl ketone	P	2	0.6	25% G16	AWBW, CCL3W	70°C; FID; N$_2$
Isosorbide Concentrate, USP	Assay	P	3	0.6	NA	S9	230°C; FID; N$_2$
Isotretinoin, USP	Organic volatile impurities, Method V <467> Limit of alcohol and formamide	C	0.53	30	3-μm G43	Fused silica	40–180°C; FID; He
Labetalol Hydrochloride, USP	Diastereoisomer ratio	P	2	1.8	10% G31, 15% G18	S1AB	320°C; FID; N$_2$

Article	Test		Diam.	Length	Packing	Support	Conditions
	Organic volatile impurities, Method I <467>						
Lamivudine, USP	Limit of residual solvents	C	0.53	50	5-μm G1	Fused silica	70–20°C; FID; H$_2$
Lanolin Modified, USP	Limit of free lanolin alcohols	C	0.53	30	0.5-μm G2	Fused silica	210–280°C; FID; N$_2$
Lanolin, USP	Pesticide residues	C	0.53	30	1.5-μm G1	Fused silica	200°C; TCD; H$_2$
Lindane Cream, USP	Assay	P	2	1.8	3% G3	S1A	195°C; FID; N$_2$
Losartan Potassium, USP	Limit of cyclohexane and isopropyl alcohol	C	0.53	30	1.5-μm G27	Fused silica	50–200°C; FID; H$_2$
Mafenide Acetate for Topical Solution, USP	Content of acetic acid	C	0.53	60	0.5-μm AW G35	Fused silica	150–240°C; FID; H$_2$
Magnesium Stearate, NF	Relative content of stearic acid and palmitic acid	C	0.32	30	0.5-μm G16	Fused silica	70–240°C; FID; He
	Organic volatile impurities, Method IV <467>						
Malathion Lotion, USP	Isopropyl alcohol content	P	4	2	5% G6	S2	130°C; FID; N$_2$
Mangafodipir Trisodium, USP	Assay	P	2	1.8	1.8-μm G43	S1A	195°C; FID; N$_2$
	Limit of residual solvents	C	0.32	30		Fused, silica	50°C; FID; He
Mecamylamine Hydrochloride Tablets USP	Dissolution	C	0.53	30	1.5-μm G27	Fused, silica	150°C; FID; He
Mecamylamine Hydrochloride, USP	Limit of residual solvents dual column	C	0.53, 0.53	30, 25	1.0-μm G16, 0.5-μm G1	Fused, silica	50–210°C; FID; N$_2$
	Assay and related compounds	C	0.53	30	1.5-μm G37	Fused, silica	120–250°C; FID; N2
Menthol Lozenges, USP	Assay	C	0.53	30	1-μm G16	Fused, silica	125°C; FID; He

(Continued)

TABLE 2.15 (*continued*)

Name of Monograph	Analyte	Packed (P) or Capillary (C)	Column i.d. (mm)	Column Length (m)	Liquid Phase (%)[b]	Support	Column Temperature; Detector; Carrier Gas
Meperidine Hydrochloride, USP	Purity	P	2	2	10% G3	S1A	190°C; FID; He
Mepivacaine Hydrochloride, USP	Purity	P	4	1.8	3% G19	S1A	230°C; FID; He
Meradimate, USP	Assay	C	0.32	25	0.1-μm G1	Fused, silica	60–240°C; FID; He
Meropenem, USP	Limit of acetone	P	3	2		S2	150°C, FID, N₂
Mesalamine, USP	Aniline, 2-aminophenol, and 4-aminophenol)	C	0.53	10	2.65-μm G47	Fused, silica	70–150°C; FID; He
Metaproterenol Sulfate, USP	Isopropyl alcohol and methanol	P	2	2	0.1% G25	S7	40–200°C; FID; He
Methadone Hydrochloride Oral Solution, USP	Organic volatile impurities, Method IV < 467 > Alcohol content, Method II < 611 >						
Methohexital Sodium for Injection, USP	Assay	P	4	1.2	3% G10	S1AB	230°C; FID; He
Methoxyflurane, USP	Assay	P	4	3	G11	S1A SS	110°C; TCD; He
Methyl Benzylidene Camphor, USP	Assay, related compounds	C	0.32	30	0.25 -μm G1	Fused-silica	100–300°C; FID; He
Methylene Chloride, NF	Assay	P	4	1.8	15% G18	S1C	60°C; TCD; He
Mibolerone Oral Solution, USP	Assay	P	3	61 cm	1% G6	S1AB	175°C; FID; He

Miconazole Nitrate Topical Powder, USP	Assay	P	2	1.2	3% G32	S1A	250°C; FID; He
Mitoxantrone Hydrochloride, USP	Alcohol	P	2	3	20% G1, 0.1% G39	S1A	50–140°C; FID; He
Moricizine Hydrochloride, USP	Limit of alcohol (C_2H_5OH)	P	2	1.8	NA	S2	150°C; FID; He
	Organic volatile impurities, Method IV <467>						
Myristyl Alcohol NF	Assay, container headspace content	P	3	2	10% G2	S1A	205°C; FID; He
	Organic volatile impurities, Method V <467>						
Naftifine Hydrochloride Gel, USP	Content of alcohol	P	3.2	1.5	NA	S3	170°C; FID; N2
Naltrexone Hydrochloride, USP	Limit of total solvents	P	4	1.8	NA	S3	170°C; FID; He/N2
Nicotine, USP	Chromatographic purity	C	0.53	30	1.5-μm G1	Fused, silica	50–250°C; FID; He
Nonoxynol 9, USP	Free ethylene oxide	P	2.1	6.4 Ni	NA	S9	100°C; FID; He
Norgestimate, USP	Limit of residual solvents—	C	0.53	30	1-μm G16	Fused, silica	65–160°C; FID; He
	Organic volatile impurities, Method IV <467>						
Octinoxate, USP	Assay Chromatographic Purity	C	0.32	25	G1	Fused, silica	80–300°C; FID; He
Octisalate, USP	Assay chromatographic purity	C	0.32	25	0.1-μm G1	Fused-silica	60–240°C; FID; N2

(Continued)

TABLE 2.15 (*continued*)

Name of Monograph	Analyte	Packed (P) or Capillary (C)	Column i.d. (mm)	Column Length (m)	Liquid Phase (%)[b]	Support	Column Temperature; Detector; Carrier Gas
Octocrylene, USP	Assay chromatographic purity	C	0.32	60	0.25-μm G1	Fused, silica	80–280°C; FID; He
Octyldodecanol, NF	Assay Organic volatile impurities, Method V < 467 >	P	2	2	3% G2	S1A	80–300°C; FID; N₂
Ofloxacin, USP	Limit of methanol and ethanol	C	0.53	30	3-μm G43	Fused, silica	35–200°C; FID; He
Oil- and Water-Soluble Vitamins with Minerals Oral Solution, Dietary Supplement	Alcohol content, Method I < 611 >						
Oxandrolone Tablets, USP	Dissolution	C	0.53	30	0.5-μm G27	Fused, silica	190–320°C; FID; He
Oxandrolone Tablets, USP	Assay	P	4	2	3% methylsilicone oil	AWBW and water-washed, flux-calcined, silanized siliceous earth	250°C; FID; He
Oxycodone Hydrochloride Oral Solution, USP	Assay same as Oxandrolone Tablets, USP						

(Continued)

Name	Test	Type			Phase	Support	Conditions
Oxycodone Hydrochloride, USP	Alcohol content, Method II 611						
Padimate O, USP	Limit of alcohol (C2H5OH)	P	4	1.8	NA	S3	150C, FID, He/N2
Paroxetine Hydrochloride, USP	Chromatographic purity	P	3	1.8, SS	10% G9	S1A	150–250°C; FID; He
	Organic volatile impurities, Method V <467>						
Penicillin G Procaine, Dihydrostreptomycin Sulfate, Chlorpheniramine Maleate, and Dexamethasone Injectable Suspension, USP	Assay for chlorpheniramine maleate	P	4	1.8, glass	1.2% G16, 0.5% KOH	S1A	180°C; FID; N_2
Perflubron, USP	Assay, chromatographic impurity	C	0.25	60	1-µm G2	Fused silica	35–185°C; FID; H_2
Perflutren Protein-Type A Microspheres for Injection USP	Assay, container headspace content	C	0.53	25	Al_2O_3/KCl coat	Porous fused silica	65°C; TCD; He
Perflutren Protein-Type A Microspheres Injectable Suspension USP	Assay, container headspace content	C	0.53	25	Al_2O_3/KCl coat	Porous fused silica	65°C; TCD; He

TABLE 2.15 (*continued*)

Name of Monograph	Analyte	Packed (P) or Capillary (C)	Column i.d. (mm)	Column Length (m)	Liquid Phase (%)[b]	Support	Column Temperature; Detector; Carrier Gas
Phendimetrazine Tartrate, USP	L-Erythro isomer	C	0.25	25	0.4-μm G1	Fused, silica	140°C; FID; H$_2$
Phendimetrazine Tartrate, USP	Organic volatile impurities, Method I < 467 >						
Phenoxyethanol, NF	Chromatographic Purity Assay	C	0.32	10	5-μm G27	Fused silica	80–260°C; FID; He
Phenyltoloxamine Citrate, USP	Related compounds	C	0.32	25	0.45-μm G27	Fused, silica	190–240°C; FID; He
	Organic volatile impurities, Method I < 467 >						
Phenytoin Sodium Injection, USP	Alcohol and propylene glycol content	P	2	1.8	NA	S3	140–190°C; FID; He
Poloxamer NF	Limit of free ethylene oxide, propylene oxide, and 1,4-dioxane	C	0.32	50	5 G27	Fused silica	70–240°C; FID; He
	Limit of free ethylene oxide, propylene oxide, and 1,4-dioxane	C	0.32	50	5-μm G27	Fused silica	70–240°C; FID; He
	Organic volatile impurities, Method V < 467 >						
Polycarbophil, USP	Limit of ethyl acetate	P	2	10 ft	1% G25	S12	160°C; FID; He
	Organic volatile impurities, Method IV < 467 >						

Article	Test	C/P			Packing	Column	Conditions
Polyethylene Glycol Monomethyl Ether, NF	Free ethylene oxide and 1,4-dioxane	C	0.32	50	5-μm G27	Fused silica	70–240°C; FID; He
	Limit of ethylene glycol and diethylene glycol	3	1.0		S2		200°C; FID; N2
	Limit of 2-methoxyethanol	C	0.053	15	1-μm G16	Fused silica	50–250°C; FID; He
Polyethylene Glycol, NF	Limit of free ethylene oxide and 1,4-dioxane	C	0.32	50	5-μm G27	Fused silica	70–240°C; FID; He
	Limit of ethylene glycol and diethylene glycol MW <450 > Da	P	3	1.5	12% G13	S1NS	140°C; FID; N2
	Organic volatile impurities, Method IV <467 >						
Polyethylene Oxide, NF	Limit of free ethylene oxide	C	0.53	10	20μm G45	Fused silica	70–200°C; FID; He
	Organic volatile impurities, Method I <467 >	C	0.53	30	5-μm G27	Fused silica	35–260°C; FID; He
Polyoxyl Ethyl Ether, USP	Limit of free ethylene oxide and dioxane	C	0.32	30	1.0μm G1	Fused, silica	50–230°C; FID; He
Primidone Oral Suspension, USP	Assay	P	4.0	120	10% G3	S1AB	260°C, FID; He
Primidone Tablets, USP	Assay	P	4.0	120	10% G3	S1AB	260°C; FID; He
Procyclidine Hydrochloride, USP	Related compounds	P	2	1	10% PEG 20K, 2% KOH	S1A	180°C; FID; He
Propofol, USP	Organic volatile impurities, Method V <467 >						
	Assay	C	053	30	1.2-μm G16	Fused silica	145–200°C; FID; He

(Continued)

TABLE 2.15 (*continued*)

Name of Monograph	Analyte	Packed (P) or Capillary (C)	Column i.d. (mm)	Column Length (m)	Liquid Phase (%)[b]	Support	Column Temperature; Detector; Carrier Gas
Propoxyphene Hydrochloride, Aspirin, and Caffeine Capsules, USP	Dissolution and assay for propoxyphene hydrochloride and caffeine	P	3	60	3% methyl phenyl silicone	Siliceous earth	175°C; FID; N$_2$
Propoxyphene Napsylate and Aspirin Tablets, USP	Assay for propoxyphene napsylate	P	3	120	3% G3	S1A	175°C; FID; N$_2$
	Dissolution and assay for propoxyphene hydrochloride and caffeine	See method for capsules					
Propylene Glycol, USP	Organic volatile impurities, Method IV < 467 >	P	4	1	5% G16	S5	120–200°C; TCD; He
Psyllium Hemicellulose, USP	Limit of alcohol	C	053	30	3.0-µm G43	Fused silica	40–230°C; FID; He
Pygeum Extract, Dietary Supplement	Content of sterols	C	0.32	30	0.25-µm G27	Fused silica	250–320°C; FID; He
Quinapril Hydrochloride, USP	Limit of residual solvents	C	0.53	30	1.0-µm G16	Fused silica	35–150°C; FID; He
Resorcinol and Sulfur Lotion	Alcohol content < 611 >						
Resorcinol and Sulfur Topical Suspension USP	Alcohol content < 611 > Alcohol content < 611 >						

Rimantadine Hydrochloride, USP	Limit of toluene	P	2	2	NA	S1A	200°C; FID; N₂
	Assay	P	4	1.8, glass	3% G19	S1A	160°C; FID; N₂
Saccharin, NF	Toluenesulfonamides	P	3.2	1.8, glass	10% G3	S1AB	210°C; FID; He
	Organic volatile impurities, Method V < 467 >						
Salsalate, USP	Limit of dimethylaniline	C	0.53	30	1.0-μm G42	Fused silica	105°C; FID; He
	Isopropyl, ethyl, and methyl salicylates	C	0.53	30	1.0-μm G42	Fused silica	120°C; FID; He
	Organic volatile impurities, Method V < 467 >						
Saw Palmetto Extract, Dietary Supplement	See above	C	0.2	25	0.33-μm G1	Fused silica	200–300°C; FID: He
	Organic volatile impurities, Method IV < 467 >						
	Alcohol content, Method II 611						
Saw Palmetto, Dietary Supplement	Assay not less than 3.0% of oleic and 2.0% of lauric acid; 1.2% of myristic acid; 1.0% of palmitic acid; 0.4% of linoleic acid; 0.2% each of caproic acid, caprylic acid, and capric acid; 0.1% of stearic acid; 0.5% of linolenic acid; and 0.01% of palmitoleic acid. Pesticide residues < 561 >	C	0.25	30	0.25-μm G16	Fused silica	120–220°C; FID; He

(Continued)

TABLE 2.15 (*continued*)

Name of Monograph	Analyte	Packed (P) or Capillary (C)	Column i.d. (mm)	Column Length (m)	Liquid Phase (%)[b]	Support	Column Temperature; Detector; Carrier Gas
Scopolamine Hydrobromide Injection, USP	Assay	P	2	1.8, glass	3% G3	S1AB	225°C; FID; N$_2$
Secobarbital Sodium and Amobarbital Sodium Capsules, USP	Assay	P	3.5	0.6, glass	3% G10	S1AB	175°C; FID; He
Sodium Caprylate, NF	Chromatographic purity	C	0.25	30	0.25-μm G25	Fused-silica	100–220°C; FID; He
Sodium Cetostearyl Sulfate, NF	Assay	C	0.25	25	G2	Fused silica	150–250°C; FID; N$_2$
	Limit of methanol, isopropyl alcohol, and acetone Organic volatile impurities, Method IV < 467 >	P	2	1.8, glass	5% G16	S12	70–180°C; FID; He
Spectinomycin Hydrochloride, USP	Assay	P	3	60cm, glass	5% G27	S1AB	190°C, FID; He
Stearic Acid, NF	Assay	P	3	1.5, glass	15% G4	S1A	165°C; FID; He
Stearyl Alcohol, NF	Assay	P	3	2, glass	10% G2	S1A	205°C; FID; He
Stinging Nettle, Dietary Supplement	Content of β sitosterol	C	0.20	25	0.35-μm G2	Fused silica	300°C; FID; He
	Pesticide residues < 561 >		0.32	30	0.25-μm G1	Fused silica	50–250°C; AFID; He
Sucralfate, USP	Limit of pyridine and 2-methylpyridine	C	0.53	10	2.65-μm G27	Fused silica	50°C; FID; He

Substance	Test						Conditions
Sucralose, NF	Organic volatile impurities, Method IV <467>						
	Limit of methanol	P	4	2, glass	NA	S6	150°C; FID; He
	Limit of acetone	P	4	1.83, glass	NA	S2	175°C; FID; N$_2$
Sufentanil Citrate, USP							
Sunflower Oil, NF	Fatty acid composition-	C	0.25	30	0.25-μm G5	Fused silica	120°C240°C; FID; H$_2$
Tamoxifen Citrate, USP	Related impurities	P	4	1, glass	5% G17	S1AB	260°C; FID; He
	Organic volatile impurities, Method V <467>						
Terazosin Hydrochloride, USP	Limit of tetrahydro-2-furancarboxylic acid	C	053	10	1.2-μm G25	Fused silica	170°C; FID; He
Terpin Hydrate and Codeine Elixir, USP	Alcohol content, Method II <611>						
	Terpin assay	P	3.5	1.2, glass	6% G1	S1A	120°C; FID; N$_2$
	Assay for codeine	P	3.5	1.2, glass	6% G1	S1A	230°C; FID; N$_2$
Terpin Hydrate and Codeine Oral Solution, USP	Alcohol content, Method II <611>						
	Assay for terpin hydrate	P	3.5	1.2, glass	6% G1	S1A	120°C; FID; N$_2$
	Assay for codeine	P	3.5	1.2, glass	6% G1	S1A	230°C; FID; N$_2$
Testosterone Cypionate, USP	Assay	P	3	1.2, glass	1% G6	S1AB	260°C; FID; He
Tetracaine and Menthol Ointment, USP	Organic volatile impurities, Method V <467>						
	Assay for menthol	P	2	1.8	10% G16	S1AB	170°C; FID; He

(Continued)

TABLE 2.15 (continued)

Name of Monograph	Analyte	Packed (P) or Capillary (C)	Column i.d. (mm)	Column Length (m)	Liquid Phase (%)[b]	Support	Column Temperature; Detector; Carrier Gas
Theophylline and Guaifenesin Oral Solution, USP	Alcohol content (if present), Method II < 611 >						
Tiamulin Fumarate, USP	Limit of residual solvents	C	0.53	30	0.5-μm G16	Fused silica	75°C; FID; He
Tiamulin, USP	Limit of alcohol toluene	C	0.53	30	1.0-μm G16	Fused silica	50–150°C; FID; He
Tiletamine and Zolazepam for Injection. USP	Assay for tiletamine and zolazepam	P	2	1.24	3% G2	S1AB	150–230°C; FID; He
Tilmicosin Injection	Content of propylene glycol	C	0.53	15	1-μm G16	Fused silica	100°C; FID; He
Tocopherols Excipient, NF	Assay	P	4	2	2–5% G2	S1AB	255°C; FID; He/N$_2$
Tolcapone, USP	Organic volatile impurities, Method IV < 467 >; Limit of residual solvents, dual column	C	0.53, 0.53	30, 5	3.0-μm G43 3-μm G3	Fused silica	35°C; FID; He
Triazolam	Assay	P	3	120cm	3% G6	S1AB	240°C; FID; He
Triazolam, USP	Chromatographic purity	P	3	120cm, glass	3% G6	S1AB	240°C; FID; He
Tributyl Citrate, NF	Assay	C	0.32	30	0.5-μm G42	Fused silica	80–220°C; FID; He
Triclosan, USP	Assay	C	0.53	15	G3	Fused silica	34–200°C; FID; He
Triclosan, USP	Limit of 2,3,7,8-tetrachlorodibenzo-p-dioxin and 2,3,7,8-tetrachlorodibenzofuran	C	0.25	60	G48	Fused silica	80–270°C; MS; He

Trientine Hydrochloride, USP	Organic volatile impurities, Method I < 467 >						
Triethyl Citrate, NF	Assay	C	0.32	30	0.5-μm G42	Fused silica	80–220°C; FID; He
Tyloxapol, USP	Limit of ethylene oxide	P	2	1.8, glass	5% G16	S12	50–200°C; FID; He
Tyloxapol, USP	Organic volatile impurities, Method I < 467 >						
Urea C13 USP		C	0.25	15	0.1-mm G47	Silica	200°C; MS; He
Valproic Acid Capsules, USP	Assay dissolution	P	2	1.8, glass	10% G34	S1A	150°C; FID; He
Valproic Acid, USP	Chromatographic purity	C	0.32	60	0.3-μm G25	Fused silica	145–190°C; FID; He
	Assay	P	2	1.8, glass	10% G34	S1A	175°C; FID; He
	Organic volatile impurities, Method V < 467 >						
Valrubicin, USP	Limit of residual solvents	C	0.32	30	0.5-μm G2	Fused silica	220°C; FID; He
Vitamin E, USP	Assay for α tocopheryl acetate	P	2	2, glass	2–5% G2	S1AB	255°C; FID; He/N$_2$
	Organic volatile impurities, Method IV < 467 >						
Vitamin E Polyethylene Glycol Succinate, NF	Assay for α-tocopherol	C	0.25	15	0.25-μm G27	Fused silica	260–340°C; FID; He
	Organic volatile impurities, Method I < 467 >						
Warfarin Sodium, USP	Isopropyl alcohol content (crystalline clathrate form)	P	4	1.8	NA	S2	140°C; FID; N$_2$
	Organic volatile impurities, Method I < 467 >						

(Continued)

TABLE 2.15 (*continued*)

Name of Monograph	Analyte	Packed (P) or Capillary (C)	Column i.d. (mm)	Column Length (m)	Liquid Phase (%)[b]	Support	Column Temperature; Detector; Carrier Gas
Water O15 Injection, USP	Radiochemical purity	C	0.53	30	G16 film	Fused silica	40°C, TCD and radioactivity; He
Xanthan Gum, NF	Limit of isopropyl alcohol Organic volatile impurities, Method IV < 467 >	P	3.2	1.8, SS	NA	S3	165°C; FID; He
Xylazine, USP	Limit of acetone and isopropyl alcohol	P	2	1.8	0.1% G25	S7	30–220°C; FID; He
Xylitol, NF	Assay Organic volatile impurities, Method I < 467 >	C	0.25	30	0.25-μm G46	Fused silica	170–270°C; FID; He
Xylose, USP	Identification Organic volatile impurities, Method I < 467 >	P	3	1.8	10% G2	S1A	170°C; FID; N$_2$
Zileuton, USP	Organic volatile impurities, Method IV < 467 > Limit of dioxane	P	2	1.8	NA	S10	140°C; FID; He/N$_2$
Reagents	Controls are not tabulated						

[a] In January 2007 many monographs will be required to meet residual solvents < 467 >.
[b] NA, not applicable.

Typical Nu-Chek-Prep ester mixtures useful in this test include Nu-Chek 17A and Nu-Chek 19A.

Nu-Chek mixture 17A has the following composition:

Percentage Fatty Acid Ester	Carbon-Chain Length	No. of Double Bonds
1.0 methyl myristate	14	0
4.0 methyl palmitate	16	0
3.0 methyl stearate	18	0
3.0 methyl arachidate	20	0
3.0 methyl behenate	22	0
3.0 methyl lignocerate	24	0
45.0 methyl oleate	18	1
15.0 methyl linoleate	18	2
3.0 methyl linolenate	18	3
20.0 methyl erucate	22	1

Nu-Chek mixture 19A has the following composition:

Percentage Fatty Acid Ester	Carbon-Chain Length	No. of Double Bonds
7.0 methyl caprylate	8	0
5.0 methyl caprylate	10	0
48.0 methyl laurate	12	0
15.0 methyl myristate	14	0
7.0 methyl palmitate	16	0
3.0 methyl stearate	18	0
12.0 methyl oleate	18	1
3.0 methyl linoleate	18	2

(Continued)

89

Chromatographic Conditions for Method: VI

Chromatographic Conditions, USP Column

Designation	Column Size	Column Temperature
A: S3	3mm × 2m	190°C
B: S2	3mm × 2.1m	160°C
C: G16	0.53mm × 30m	40°C
D: G39	3mm × 2m	65°C
E: G16	3mm × 2m	70°C
F: S4	2mm × 2.5m	Hold 120°C (35 min.) gradient 120–200(2/ min.) Hold 20 min.
H: G14	2mm × 2.5m	Hold 45°C (3 min.) gradient 45–120 (8/ min.) Hold 15 min.
I: G27	0.53mm × 30m	Hold 35°C (5 min.) 35–175°C (8/ min.) 175–260(35/ min.)Hold 16 min.
J: G16	0.33mm × 30m	Hold 50°C (20 min.) 50–165°C (6/ min.)

REFERENCES

1. W. A. George, Ph.D. dissertation, University of Massachasetts–Lowell, 1986.
2. 2005–2006 Catalogue of Supelco Chromatography Products, Supelco, Bellefonte, PA.
3. *Diatomite Supports for Gas–Liquid Chromatography*, Johns-Manville Corporation, Ken-Caryl Ranch, Denver, CO, 1981.
4. W. R. Supina, *The Packed Column in Gas Chromatography,* Supelco, Bellefonte, PA, 1974.
5. J. J. Kirkland, *Anal. Chem.*,**35**, 2003 (1963).
6. Publication FF-202A, Johns-Manville Corporation, Ken-Caryl Ranch, Denver, CO, 1980.
7. *Chromosorb Century Series Porous Polymer Supports,* Johns-Manville Corporation, Ken-Caryl Ranch, Denver, CO, 1980.
8. Catalogue of Chromatography Chemicals and Accessories, Foxboro/Analabs, North Haven, CT, K-23B.
9. D. M. Ottenstein, *J. Chromatogr. Sci.*, **11**, 136 (1973).
10. D. M. Ottenstein, in *Advances in Chromatography,* Vol. III, J. C. Giddings and R. A. Keller (Eds.), Marcel Dekker, New York, 1996, p. 137.
11. W. W. Hanneman, *J. Gas Chromatogr.,* **1**, 18 (1963).
12. J. A. Favre and L. R. Kallenbach, *Anal. Chem.,* **36**, 63 (1964).
13. P. W. Salomon, *Anal. Chem.,* **36**, 476 (1964).
14. L. B. Rogers and A. G. Altenau, *Anal. Chem.,* **35,** 915 (1963).
15. L. B. Rogers and A. G. Altenau, *Anal. Chem.,* **36**, 1726 (1964).
16. L. B. Rogers and A. G. Altenau, *Anal. Chem.*,**37,** 1432 (1965).
17. R. L. Grob, G. W. Weinert, and J. W. Drelich, *J. Chromatogr.,* **30**, 305 (1967).
18. D. J. Brookman and D. T. Sawyer, *Anal. Chem.,* **40,** 106 (1968).
19. D. J. Brookman and D. T Sawyer, *Anal. Chem.,* **40,** 1368 (1968).
20. D. J. Brookman and D. T. Sawyer, *Anal. Chem.,* **40,** 1847 (1968).
21. D. J. Brookman and D. T. Sawyer, *Anal. Chem.,* **40,** 2013 (1968).
22. G. L. Hargrove and D. T. Sawyer, *Anal. Chem.,* **40,** 409 (1968).
23. R. L. Grob, R. J. Gondek, and T. A. Scales, *J. Chromatogr.,* **53,** 477 (1970).
24. W. D. Garner and S. J. Veal, *J. Chem. Soc.,* 1487 (1935).
25. W. D. Garner, *J. Chem. Soc.,* 1239 (1947).
26. J. J. Kipling and D. B. Peakall, *J. Chem. Soc.,* 332 (1957).
27. A. L. Dent and R. J. Kohes, *J. Am. Chem. Soc.,* **92,** 1092 (1970).
28. E. Gil-Av, J. Herling, and J. Shabtai, *J. Chromatogr.,* **1,** 508 (1958).
29. J. Herling, J. Shabtai, and E. Gil-Av, *J. Chromatogr.,* **8,** 349 (1962).
30. J. Shabtai, J. Herling, and E. Gil-Av, *J. Chromatogr.,* **2,** 406 (1959).
31. J. Shabtai, J. Herling, and E. Gil-Av, *J. Chromatogr.,* **11,** 32 (1963).
32. R. L. Grob and E. J. McGonigle, *J. Chromatogr.,* **59,** 13 (1971).
33. E. J. McGonigle and R. L. Grob, *J. Chromatogr.,* **101,** 39 (1974).
34. M. J. O'Brien and R. L. Grob, *J. Chromatogr.,* **155,** 129 (1978).
35. N. J. Harrick, *Internal Reflection Spectroscopy,* Interscience, New York, 1967.

36. R. L. Grob, M. A. Kaiser, and M. J. O'Brien, *Am. Lab.,* **7,** 33 (1975).

37. R. L. Grob, M. A. Kaiser, and M. J. O'Brien, *Am. Lab.,* **9,** 13 (1975).

38. M. J. E. Golay, in *Gas Chromatography, 1957* (East Lansing Symposium), V. J. Coates, H. J. Noebels, and I. S. Fagerson, Eds., Academic Press, New York, 1958, pp. 1–13.

39. R. Dandeneau and E. H. Zerenner, *J. High Resolut. Chromatogr. Chromatogr. Commun.,* **2,** 351 (1971).

40. R. Dandeneau and E. H. Zerenner, *Proc. 3rd International Symposium on Glass Capillary Gas Chromatography,* Hindelang, Germany, 1979, pp. 81–97.

41. T. I. Wishousky, R. L. Grob, and A. G. Zacchei, *J. Chromatogr.,* **249,** 1 (1982).

42. T. I. Wishousky, R. L. Grob, and A. G. Zacchei, *J. Chromatogr.,* **249,** 155 (1982).

43. O. L. Hollis, *Anal. Chem.,* **38,** 309 (1966).

44. O. L. Hollis and W. V. Hayes, *J. Gas Chromatogr.,* **4,** 235 (1966).

45. G. E. Pollack, D. O'Hara, and O. L. Hollis, *J. Chromatogr. Sci.,* **22,** 343 (1984).

46. W. A. Supina, in *Modern Practice of Gas Chromatography,* 2nd ed., R. L. Grob (Ed.), Wiley, New York, 1985, p. 124.

47. R. Kaiser, *Chromatographia,* **1,** 199 (1968).

48. A. Di Corcia and R. Samperi, *Anal. Chem.,* **46,** 140 (1974).

49. A. DiCorcia and R. Samperi, *Anal. Chem.,* **51,** 776 (1979).

50. A. DiCorcia, R. Samperi, and C. Severini, *J. Chromatogr.,* **170,** 325 (1980).

51. R. J. Bartram, in *Modern Practice of Gas Chromatography,* 4th ed., R. L. Grob and E. F. Barry (Eds.), Wiley, Hoboken, NJ, 2004.

52. R. V. Golovnya and T. A. Mishharina, *J. High Resolut. Chromatogr.,* **3,** 4 (1980).

53. C. F. Poole and S. K. Poole, *Chem. Rev.,* **89,** 377 (1989).

54. H. Lamperczyk, *Chromatographia,* **20,** 283 (1985).

55. W. O. McReynolds, *J. Chromatogr. Sci.,* **8,** 685 (1970).

56. L. Rohrschneider, *J. Chromatogr.,* **22,** 6 (1966).

57. L. Rohrschneider, in *Advances in Chromatography,* Vol. IV, J. C. Giddings and R. A. Keller (Eds.), Marcel Dekker, New York, 1967, p. 333.

58. E. Kovats, *Helv. Chim. Acta,* **41,** 1915 (1958).

59. W. O. McReynolds, *Gas Chromatographic Retention Data,* Preston Technical Abstracts, Evanston, IL, 1956.

60. L. Ettre, *Anal. Chem.,* **36,** 31A (1964).

61. 2005 Catalogue of Chromatography Products, Restek Corporation, Bellefonte, PA.

62. J. K. Haken, *J. Chromatogr.,* **300,** 1 (1984).

63. J. J. Leary, J. B. Justice, S. Tsuge, S. R. Lowry, and T. L. Isenhour, *J. Chromatogr. Sci.,* **11,** 201 (1973).

64. R. Delley and K. Friedrich, *Chromatographia,* **10,** 593 (1977).

65. S. Hawkes, D. Grossman, A. Hartkopf, T. Isenhour, J. Leary, and J. Parcher, *J. Chromatogr. Sci.,* **13,** 115 (1975).

66. R. L. Grob, in *Modern Practice of Gas Chromatography,* 4th ed., R. L. Grob and E. F. Barry (Eds.), Wiley, Hoboken, NJ, 2004.

67. J. H. Purnell, *Nature,* **184,** 2009 (1959).

68. J. H. Purnell, *J. Chem. Soc.,* **54,** 1268 (1960).

69. R. Kaiser, *Z. Anal. Chem.,* **189,** 11 (1962).

70. L. A. Colon and L. J. Baird, in *Modern Practice of Gas Chromatography,* 4th ed., R. L. Grob and E. F. Barry (Eds.), Wiley, Hoboken, NJ, 2004.

71. *The Packed Column Application Guide,* Bulletin 890, Supelco, Bellefonte, PA, 1995.

3 Capillary Column Gas Chromatography

3.1 INTRODUCTION

A *capillary column*, also referred to as an *open-tubular column* because of its open flow path, offers a number of advantages over a packed column. These merits include vastly improved separations with higher resolution, reduced time of analysis, smaller sample size requirements, and often, higher sensitivities. The arrival of fused silica as a capillary column material had a major impact on capillary gas chromatography, and, in fact, markedly changed the practice of gas chromatography. In 1979, the number of sales of gas chromatographs with capillary capability was less than 10%; this figure increased to 60% by 1989 (1) and has escalated since then as further developments in capillary column technology and instrumentation continue to be made. In this chapter, theoretical and practical considerations of the capillary column are discussed.

Significance and Impact of Capillary Gas Chromatography

Marcel Golay is credited with the discovery of the capillary column. In 1957, Golay presented the first theoretical treatment of capillary column performance when he illustrated that a long length of capillary tubing with a thin coating of stationary phase coating the narrow inner diameter of the tube offered a tremendous improvement in resolving power compared over that of a conventional packed column (2). Such a column is also often referred to as a *wall-coated open-tubular* (WCOT) *column*. The high permeability or low resistance of a capillary column to carrier gas flow enables a very lengthy column to generate a large number of theoretical plates.

In contemporary practice, separations of high resolution are attainable by capillary GC, as illustrated in the chromatogram shown in Figure 3.1, which was generated with a conventional fused-silica capillary column. An exploded view of this chromatogram of a gasoline-contaminated Jet A fuel mixture with a data acquisition system indicates the presence of over 525 chromatographic peaks. Because of its separation power, capillary gas chromatography has become synonymous

Columns for Gas Chromatography: Performance and Selection, by Eugene F. Barry and Robert L. Grob
Copyright © 2007 John Wiley & Sons, Inc.

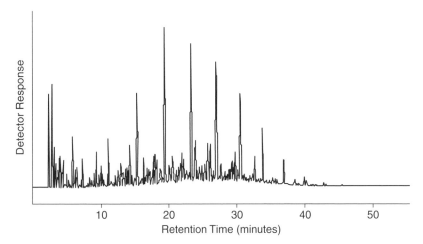

Figure 3.1   Chromatogram of a sample of Jet A fuel contaminated with gasoline on 30 m × 0.25 mm i.d. HP-1 (0.25-μm film). Column temperature conditions 30°C (5 min), 2°C/min to 250°C; split injection (100 : 1), Detector FID. [From E. F. Barry, in *Modern Practice of Gas Chromatography*, 4th ed., R. L Grob and E. F Barry (eds.), copyright © 2004 John Wiley & Sons, Inc., with permission.]

with the term *high-resolution gas chromatography*. In many cases, the high resolution afforded by capillary columns permits a direct comparison of complex samples, as illustrated in the chromatograms of two mineral spirit samples in Figure 3.2.

Chronology of Achievements in Capillary Gas Chromatography

The first column materials employed in the developmental stage of the technique were fabricated from plastic materials (Tygon and nylon) and metal (aluminum, nickel, copper, stainless steel, and even gold). Plastic capillaries, being thermoplastic in nature, had temperature limitations, whereas metallic capillary columns had the disadvantage of catalytic activity. Rugged flexible stainless steel columns rapidly became state of the art and were widely used for many applications, primarily for petroleum analyses. The reactive metallic surface proved to be too unfavorable in the analysis of polar and catalytically sensitive species. In addition, only the split-injection mode was available for quite some time for the introduction of the small quantities of sample dictated by a thin film of stationary phase within the column. Stainless steel capillary columns are shown in Figure 1.2.

As the surface chemistry of glass was gradually studied and understood, capillaries made of borosilicate and soda–lime glass became popular in the 1970s and replaced metal capillary columns (3). Here the metal oxide content and presence of silanol groups on the glass surface necessitated carefully controlled deactivation and coating procedures, but separations obtained on glass capillaries were clearly

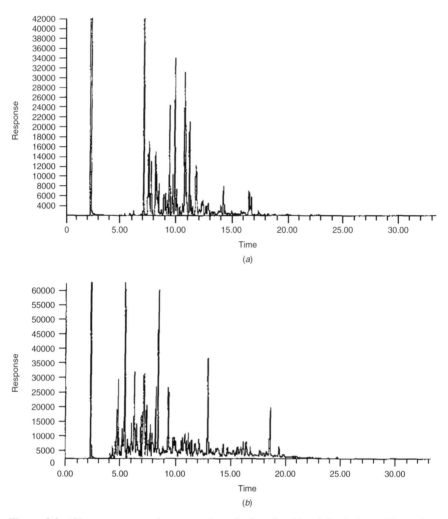

Figure 3.2 Chromatograms of two samples of mineral spirits: (*a*) odorless; (*b*) regular. Conditions: 30 m × 0.25 mm i.d. DB-1 with 0.25-μm film thickness, 40°C at 4°C/min to 150°C, split injection, 30 cm/s He, FID.

superior to those obtained with metal capillaries. The fragility of glass often proved to be problematic, requiring restraightening of a capillary end upon breakage with a minitorch followed by a recoating of the straightened portion with a solution of stationary phase to deactivate the straightened segment. Frustration was part of this endeavor and patience was also helpful! Today, equivalent or superior separations with a fused-silica capillary column (Figure 1.2) can be generated with the additional feature of ease of use.

The most significant advancement in capillary gas chromatography occurred in 1979 when Dandeneau and Zerenner at Hewlett-Packard introduced fused silica as a column material (4,5). The subsequent emergence of fused silica as the column material of choice for high-resolution gas chromatography is responsible for the widespread use of the technique and has greatly extended the range of gas chromatography. In the next two decades, there were several other major developments in capillary gas chromatography. Instrument manufacturers responded to the impact of fused-silica columns by designing chromatographs with injection and detector systems optimized in performance for fused-silica columns. There were also concurrent advances in the area of microprocessors. Reporting integrators and fast data acquisition systems with increased sampling rates now are available to be compatible with the narrow bandwidths of capillary peaks. The stature of the capillary column has been enhanced further by continuing improvements in the performance and thermal stability of the stationary phase within the column. A column containing a cross-linked phase (e.g., a silarylene or silphenylene phase), has an extended lifetime because it has high thermal stability and can tolerate large injection aliquots of solution without redistribution of the stationary phase. Inlet discrimination was addressed with the development of on-column injection, the programmed-temperature vaporizer, electronic pressure-controlled injection, and more recently, the large-volume injector with cool-on column inlet mode and solvent vapor exit.

Since Golay's proposal of the use of the capillary column, capillary gas chromatography has exhibited spectacular growth, maturing into a powerful analytical technique. Some of the more notable achievements in capillary gas chromatography are listed in Table 3.1.

TABLE 3.1   Noteworthy Events in Capillary Gas Chromatography

Year	Achievement
1958	Theory of capillary column performance, GC Symposium in Amsterdam
1959	Sample inlet splitter
	Patent on capillary columns by Perkin-Elmer
1960	Glass drawing machine developed by Desty
1965	Efficient glass capillary columns
1975	First capillary column symposium
1978	Splitless injection
1979	Cold on-column injection
	Fused silica introduced by Hewlett-Packard
1981–1984	Deactivation procedures and cross-linked stationary phases
1981–1988	Interfacing capillary columns with spectroscopic detectors (MS, FTIR, AED)
1983	Megabore column introduced as an alternative to the packed column
1992–2005	Programmed-temperature vaporizer electronic pressure-controlled sample inlet systems; -ms grade columns, solid-phase microextraction sampling techniques, large-volume injectors, integrated guard columns, multidimensional GC, sol-gel columns, improvements in silphenylene phases, advances in GC-MS, more affordable benchtop GC-MS systems

Source: Some data from Agilent Technologies.

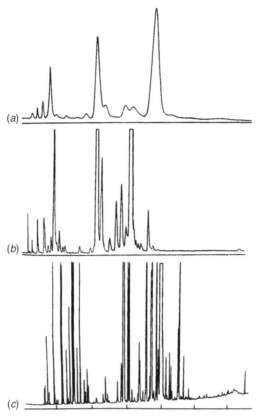

Figure 3.3 Optimized separations of peppermint oil on (*a*) 6 ft × 0.25 in i.d. packed column, (*b*) 500 ft × 0.03 in. i.d. stainless steel capillary column, and (*c*) 50 m × 0.25 mm i.d. glass capillary column. Stationary phase on each column was Carbowax 20M. [From W. Jennings, *J. Chromatogr. Sci.*, 17, 637 (1977), by permission of Preston Publications, a Division of Preston Industries, Inc.]

Comparison of Packed and Capillary Columns

Three stages in the evolution of the capillary column technology are presented in Figure 3.3: a packed column separation and two separations with a stainless steel and glass capillary column. Better resolution is evident with the capillary chromatograms because more peaks are separated and smaller peaks can be detected. The superior performance of the glass capillary column is clearly apparent in this case.

In addition to providing a separation where peaks have narrower bandwidths compared to those of a packed column counterpart, a properly prepared fused-silica capillary column, which has an inert surface (less potential for adverse adsorptive effects toward polar species), yields better peak shapes (i.e., bands are sharper with less peak tailing), which facilitates trace analysis as well as providing more

TABLE 3.2 Comparison of Wall-Coated Capillary Columns with Packed Columns

	Packed	Capillary
Length (m)	1–5	5–60
Inner diameter (mm)	2–4	0.10–0.53
Plates per meter	1000	5000
Total plates	5000	300,000
Resolution	Low	High
Flow rate (mL/min)	10–60	0.5–15
Permeability (10^7 cm^2)	1–10	10–1000
Capacity	10 μg/peak	>100 ng/peak
Liquid film thickness (μm)	10	0.1–1

Source: Data from refs. 6 and 7.

reliable quantitative and qualitative analyses. Sharp narrow bands of the trace components present in a capillary chromatogram such as that in Figures 3.1 and 3.2 have increased peak height relative to the peak of the same component at identical concentration in a packed column chromatogram, where the peak may be unresolved or disappear in the baseline noise. Moreover, due to the low carrier gas flow rate, greater detector sensitivity, stability, and signal-to-noise levels are possible with a capillary column. One drawback of the capillary column is its limited sample capacity, which requires dedicated inlet systems to introduce small quantities of sample commensurate with small amounts of stationary phase. Operational parameters of packed and capillary columns are contrasted further in Table 3.2.

The superior performance of a capillary column can be viewed further in the following manner. Because of the geometry and flow of gas through a packed bed, molecules of the same solute can take a variety of paths through the column enroute to the detector via eddy diffusion, as discussed in Chapter 2, whereas in a capillary column all flow paths have nearly equal length. The open geometry of a capillary column causes a lower pressure drop, allowing longer columns to be used. Since a packed column contains much more stationary phase, often thickly coated on an inert solid support, there are locations in the packing matrix where the stationary phase spans or spreads over to adjacent particles (Figure 2.2a). Some molecules of the same component encounter thinner regions of stationary phase, whereas other molecules have increased residence times in these thicker pools of phase, all of which create band broadening. On the other hand, a capillary column contains a relatively thin film of stationary phase coated uniformly on the inner wall of the tubing, promoting much more favorable mass transfer and less band broadening. These factors, considered collectively, are responsible for the sharp band definition and narrow retention-time distribution of molecules of a component eluting from a capillary column. Sharper band definitions also mean better detectability.

At higher column oven temperatures with increased linear velocity of carrier gas, capillary separations can be achieved that mimic those on a packed column but with a shortened time of analysis. The reduced amount of stationary phase in a capillary column imparts another advantage to the chromatographer: One observes less bleed of stationary phase from the column at elevated temperatures, and this results in less detector contamination. Theoretical considerations of the capillary column are discussed in Section 3.4.

3.2   CAPILLARY COLUMN TECHNOLOGY

Capillary Column Materials

After Desty developed a glass drawing and coiling apparatus (8), focus shifted away from metal capillary columns to fabrication of columns from more inert borosilicate and soda-lime glass. Glass is inexpensive and readily available, and glass columns could be drawn conveniently in-house with dimensions (length and inner diameter) tailored to individual needs. Investigators quickly realized that this increase in the column inertness of glass was at the expense in flexibility. With fused silica, a column could be fabricated from a material having the flexibility of stainless steel with an inner surface texture more inert than glass. Thus, fused silica quickly replaced glass as the capillary column material of choice.

Fused Silica and Other Glasses

The widespread use of fused silica as a column material for capillary GC may be attributed to the more inert inner surface texture compared to other glasses. The interested reader may refer to the book by Jennings (9) for a comparison of fused silica with other glasses, such as soda-lime, borosilicates, and lead, that have been used as column materials where the proposed structure of fused silica (Figure 3.4) may also be found; representative metal oxide concentration data are

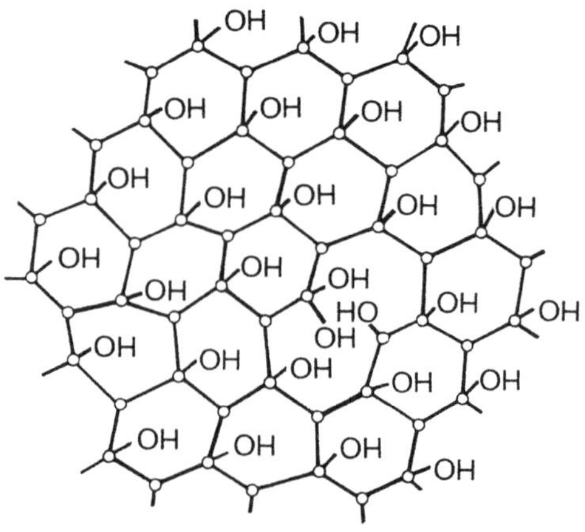

Figure 3.4   Probable structure of fused silica. (From ref. 9, with permission of Alfred Heuthig Publishers.)

TABLE 3.3   Approximate Glass Composition (Percent)[a]

Glass	SiO_2	Al_2O_3	Na_2O	K_2O	CaO	MgO	B_2O_3	PbO	BaO
Soda-lime (Kimble R6)	68	3	15	—	6	4	2	—	2
Borosilicate (Pyrex 7740)	81	2	4	—	—	—	13	—	—
Potash soda-lead (Corning 120)	56	2	4	9	—	—	—	29	—
Fused silica	100								

Source: Data from refs. 6 and 9.
[a]Less than 1 ppm total metals.

TABLE 3.4   Typical Impurities Present in Fused Silica

Trace Element	Concentration (ppb)
Aluminum	20
Calcium	<10
Chlorine	0
Copper	<8
Iron	<8
Lithium	<10
Sodium	<8
Potassium	<10
Magnesium	<10
Manganese	<5
Titanium	<10
Zirconium	<10

Source: www.sge.com/scientific tubing.

presented in Table 3.3. Column activity may be attributed to exposed silanol groups and metal ions on the surface of a glass capillary. While soda-lime borosilicate glasses, for example, have percentage levels of metal oxides, the metal content of synthetic fused silica is less than 1 ppm. Table 3.4 presents a more detailed breakdown of typical impurities found in fused silica at the ppb level. Further engineering, physical, mechanical, and testing properties of interest are presented in Table 3.5. The data in Table 3.5 are probably of little interest to the practicing chromatographer but are presented here in the interest of diversity, as a testimony to the power of the Internet and as an illustration of information that may be found at the Web site of a column manufacturer.

Although quartz tubing is available commercially, its metal oxide content (10 to 100 ppm) is considered to be too great for use in capillary gas chromatography (10). Metal oxides are Lewis acids and can serve as adsorptive sites for electron-donor species such as ketones and amines and as an active site for species with *p*-bonding

TABLE 3.5   **Physical and Mechanical Properties of Fused Silica**

Characteristic at 20°C	
Density (g/cm^3)	2.20
Hardness (micro-Vickers) (kg/mm^3)	765–800
Hardness, Moh's	7
Young's modulus (kg/cm^2)	741.32
Rigidity modulus (kg/cm^2)	316.95
Poisson's ratio	0.180
Compressibility (kg/cm^2)	11.50
Tensile strength (kg/cm^2)	500
Bending strength (kg/cm^2)	700
Torsional strength (kg/cm^2)	300
Thermal conductivity (W/m · K)	1.38

Source: SGE Web site, www.sge.com/scientific tubing.

capability (aromatics and olefins). Boron impurities in glass also act as Lewis acid sites capable of chemisorbing electron donors (9). The absence of these adsorptive sites in fused silica is responsible for its remarkable inertness and is a direct result of the synthesis of this material. However, the hydroxyl groups attached to tetravalent silicon atoms are of paramount significance because they can contribute residual column activity.

Synthetic fused silica is formed by introducing pure silicon tetrachloride into a high-temperature flame followed by reaction with the water vapor generated in the combustion (9, 11). The process can be described by the reaction

$$SiCl_4 + 2H_2O \rightarrow SiO_2 + 4HCl \tag{3.1}$$

Three distinct categories of silanol groups are present on the surface of the fused silica shown in Figure 3.4. First, there are free silanol groups, which are acidic adsorptive sites with $K_a = 1.6 \times 10^{-7}$. The surface concentration of free silanols on fused silica has been calculated to be 6.2 μmol/m^2. This type of silanol group has a direct bearing on column behavior. Geminal silanols, the situation where two hydroxyl groups are attached to the same silicon atom, are also present at a concentration of 1.6 μmol/m^2 (12). Third, there are vicinal silanol functionalities characterized by hydroxyl groups attached to adjacent silicon atoms. Here steric effects become important. For example, vicinal silanols represent a rather weak adsorptive site, but in the presence of water can be rendered active (9):

If the interatomic distance of neighboring oxygen atoms is between 2.4 and 2.8 Å, the groups are hydrogen bonded, if this distance exceeds 3.1 Å, hydrogen bonding does not occur and free silanol behavior dominates (13). Bound silanol groups can dehydrate, producing siloxane bridges under certain conditions. More detailed information on the complexities of silica surface chemistry may be found in the books by Jennings (9,11) and Lee et al. (6).

Extrusion of a Fused-Silica Capillary Column

Three steps are involved in the preparation of a fused-silica column: (1) the high-temperature extrusion of the blank capillary tubing from a preform, where the capillary receives a protective outer coating in the same process; (2) the deacti-vation of the inner surface of the column; and (3) the uniform deposition of a stationary phase of a desired film thickness on the deactivated inner surface. In this section the extrusion of fused silica is described; the procedures employed for the deactivation and coating of fused-silica capillaries are presented in the following section.

Prior to extrusion, the fused-silica preform is usually treated with dilute hydrofluoric acid to remove any imperfections and deformations present on the inner and outer surfaces and then rinsed with distilled water and followed by annealing (14). The extrusion is conducted in an elaborate, sophisticated fused-silica drawing facility (Figure 3.5); a picture of a drawing tower is presented in Figure 3.6. An overall schematic diagram of the drawing of raw fused silica is depicted in Figure 3.7. In a clean-room atmosphere, the preform is drawn vertically through a furnace maintained at approximately 2000°C. Guidance and careful control of the drawing process is achieved by focusing an infrared laser beam down the middle of the capillary in conjunction with feedback control electronic circuitry in order to maintain uniformity in the specifications of the

Figure 3.5   Facility for drawing fused-silica capillary tubing. (Courtesy of SGE Analytical Sciences.)

Figure 3.6 Fused-silica drawing tower. (Courtesy of SGE Analytical Sciences.)

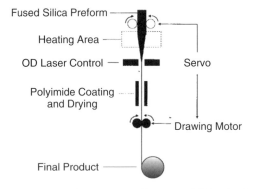

Figure 3.7 Fused-silica drawing process. (Courtesy of SGE Analytical Sciences.)

inner and outer diameters in the final product. The final product is coiled on a takeup drum (Figure 3.8).

The publications of Macomber et al. are recommended to the interested reader desiring more in-depth knowledge of the properties and process of preparing fused capillaries not only for GC but also for capillary LC, capillary electrophoresis (CE), and capillary electrochromatography (CEC) (15–21). As noted by Macomber

Figure 3.8   Takeup spool for containing fused-silica tubing after drawing.

TABLE 3.6   Fused-Silica Tubing Specifications (μm) in the Period 1990–2001[a]

Column Product	OD Specifications		ID Specifications	
(mm i.d.)	1990	2001	1990	2001
0.25	350(15)	360(10)	250(12)	250(6)
0.32	430(20)	435(10)	320(12)	320(6)
0.53	665(25)	665(15)	542(12)	536(6)

[a]Numbers in parentheses represent standard deviations.

Source: Data from ref. (17).

et al. (17), with the evolution of the GC market, suppliers of fused-silica tubing strive for greater i.d. accuracy and tighter i.d. tolerances (Table 3.6).

Drawn fused silica exhibits very high tensile strength and has excellent flexibility due to the thin wall of the capillary. However, the thin wall of the capillary is subject to corrosion upon exposure to atmospheric conditions and is extremely fragile. To eliminate degradation and increase its durability, the fused-silica tubing receives a protective outer coating, usually of polyimide, although other coating materials have been used, including silicones, gold, vitreous carbon, acrylate, polyamides, and aluminum. Polyimide, which also serves as a water barrier, is most widely used because it offers temperature stability to 400°C. The polyimide coating is also referred to as a stress buffer and is typically 15 to 20 μm in thickness. The color of polyimide seems to vary slightly from one column manufacturer to another, with no effect, however, on column performance. An excellent historical review on the story behind the technology and extrusion of fused-silica tubing has recently been published (22). Jennings prepared a fascinating recounting of his own personal perspectives on the development and commercialization of the capillary column (23).

Aluminum-Clad Fused-Silica Capillary Columns

There are number of application areas requiring columns to be operated at or above 400°C, such as the analysis of waxes, crude oils, and triglycerides. These have driven efforts to replace the polyimide outer coating with a thin layer of aluminum and extend the temperature range of capillary gas chromatography, as illustrated in the chromatograms of high-temperature capillary separations shown in Figure 3.9. An aluminum-clad capillary column has excellent heat transfer while maintaining the same flexibility and inertness of the fused-silica surface as the polyimide-coated columns. Trestianu and co-workers showed that for an alkane of high carbon number, the elution temperature on a high-temperature column is 100°C lower than with a corresponding packed column (24). It must be emphasized here that to obtain optimum column performance, the injection mode is critical. Cold on-column or programmed-temperature vaporizer injectors are recommended to avoid inlet discrimination problems for the analysis of solutes of high molecular weight with this type of capillary column. The central column in Figure 1.2 is an aluminum-clad fused-silica column.

Fused-Silica-Lined Stainless Steel Capillary Columns

A third type of protective outer coating, stainless steel, for fused silica offers an alternative to aluminum-clad fused silica for elevated column temperatures. This technology is the inverse of that for a polyimide-clad fused-silica capillary, where a layer of fused silica is deposited on the inner surface of a stainless capillary. In Figure 3.10, scanning electron micrographs are displayed which compare the rough surface of stainless steel with the smooth surface of untreated fused silica and the surface of stainless steel after a 1-μm layer of deactivated fused silica is bonded to its interior wall, termed Silcosteel, available commercially as the MXT series of columns (Restek Corporation). In Figure 3.11, applications of this column technology to peppermint oil and petroleum samples are displayed. In addition to high thermal stability, a distinguishing feature of a fused-silica-lined thin-walled stainless steel capillary column is that it can be coiled in a diameter of less than 4 in. (compared to larger diameters with polyimide-clad fused silica) without breakage, making it a very favorable column material for process control and portable gas chromatographs, where the size of a column oven, shock resistance, and ruggedness become limiting factors. Photographs of an MXT and a conventional fused-silica column are shown in Figure 3.12. The Silcosteel treatment has also been extended to passivate metal liners for split or splitless injections in conjunction with a high-temperature deactivation procedure, offering an economical alternative to standard glass liners, which are breakable.

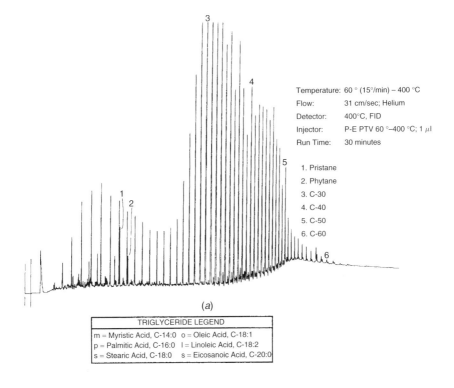

Temperature: 60 ° (15°/min) – 400 °C
Flow: 31 cm/sec; Helium
Detector: 400°C, FID
Injector: P-E PTV 60 °–400 °C; 1 μl
Run Time: 30 minutes

1. Pristane
2. Phytane
3. C-30
4. C-40
5. C-50
6. C-60

(a)

TRIGLYCERIDE LEGEND	
m = Myristic Acid, C-14:0	o = Oleic Acid, C-18:1
p = Palmitic Acid, C-16:0	l = Linoleic Acid, C-18:2
s = Stearic Acid, C-18:0	s = Eicosanoic Acid, C-20:0

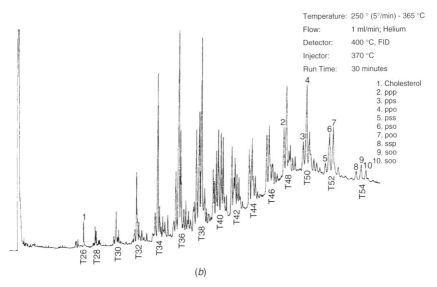

Temperature: 250 ° (5°/min) - 365 °C
Flow: 1 ml/min; Helium
Detector: 400 °C, FID
Injector: 370 °C
Run Time: 30 minutes

1. Cholesterol
2. ppp
3. pps
4. ppo
5. pss
6. pso
7. poo
8. ssp
9. soo
10. soo

(b)

Figure 3.9 Chromatogram of separation of (*a*) Canadian wax on 15 m × 0.25 mm i.d. aluminum-clad capillary column (0.1-μm film), (*b*) triglycerides on a 25 m × 0.25 mm i.d. aluminum-clad capillary column (0.1-μm film), (*c*) Upper: "good" biodiesel fuel, (*d*) lower: "bad" biodiesel fuel. (Courtesy of the Quadrex Corporation.) (*Continued*)

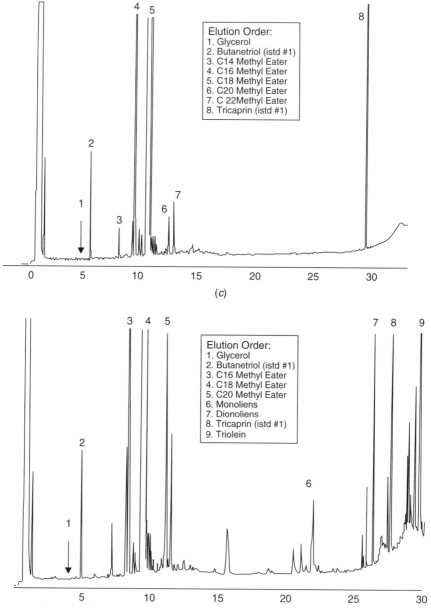

Elution Order:
1. Glycerol
2. Butanetriol (istd #1)
3. C14 Methyl Eater
4. C16 Methyl Eater
5. C18 Methyl Eater
6. C20 Methyl Eater
7. C 22Methyl Eater
8. Tricaprin (istd #1)

(c)

Elution Order:
1. Glycerol
2. Butanetriol (istd #1)
3. C16 Methyl Eater
4. C18 Methyl Eater
5. C20 Methyl Eater
6. Monoliens
7. Dionoliens
8. Tricaprin (istd #1)
9. Triolein

Column:5% Phenyl Methyl Silecone, 10-m x 0.23 mm I.D. x 0.1um film. AL-clad fused silica; Column Temperature: 50 C (1 min)-15 C/min to 180 C 2 7 c/min to 230 C @30 C /min to 380 C (10 min hold); cool on-column injection,1 uL; FID, 380 C; He at 1.8 mL/min.

(d)

Figure 3.9   (*continued*)

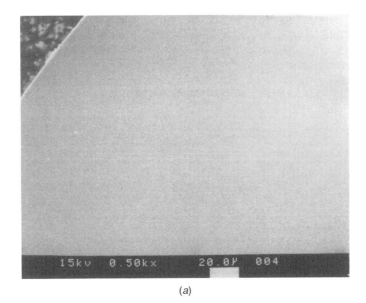

(a)

(b)

Figure 3.10   Scanning electron micrographs of (a) untreated fused silica; (b) the rough inner surface of stainless steel capillary tubing; (c) the smoother inner surface of the stainless steel capillary tubing after deposition of a thin layer of fused silica; (c) also illustrates regions where fused-silica lining was removed selectively to expose untreated stainless steel surface below. (Courtesy of the Restek Corporation.) (*Continued*)

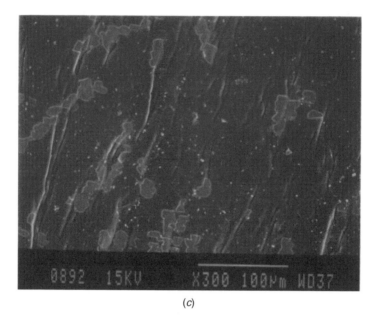

(*c*)

Figure 3.10 (*continued*)

Along the same line, an aluminum capillary lined with quartz coated with carbon black has been evaluated for the analysis of amines, Volatile Organic Compounds (VOCs), and oil products (25). In an alternative stainless steel column format (non-glass-lined stainless steel) introduced by Agilent Technologies, the metal column is deactivated and thus provides inertness similar to that of fused silica. Columns of this nature, termed DB-ProSteel series, have the same o.d. as that of a standard megabore column (0.53 mm i.d.) and require no special ferrules.

3.3 PREPARATION OF FUSED-SILICA CAPILLARY COLUMNS

Most users of a modern capillary column regard it as a sophisticated high-precision device and purchase columns from a vendor. Few give any thought to the steps involved in column preparation. Their highest priority is understandably the end result of accurate and reproducible chromatographic data that the column can provide for quality assurance/quality control protocols. In this section, deactivation and coating of a fused-silica column with stationary phase are discussed.

Silanol Deactivation Procedures

For maximum column performance, blank or raw fused-silica tubing must receive pretreatment prior to the final coating with stationary phase. The purpose of pretreatment is twofold: to cover up or deactivate active surface sites and to create a surface more wettable by the phase. The details of the procedure depend on the stationary phase to be subsequently coated, but deactivation is essential for producing a column that has a uniform film deposition along the inner wall of the capillary.

Although metal ions are not a factor with fused silica, the presence of silanol groups must still be addressed; otherwise, the column has residual surface activity. Column activity can be demonstrated in several ways. The chromatographic peak of a given solute can disappear completely, the peak can disappear partially, being diminished in size, or a peak can exhibit tailing. A chromatogram of a test mixture showing the activity of an uncoated fused-silica column is displayed in Figure 3.13a; the inherent acidity associated with surface silanol groups is responsible for the complete disappearance of the basic probe solute, 2,6-dimethylaniline.

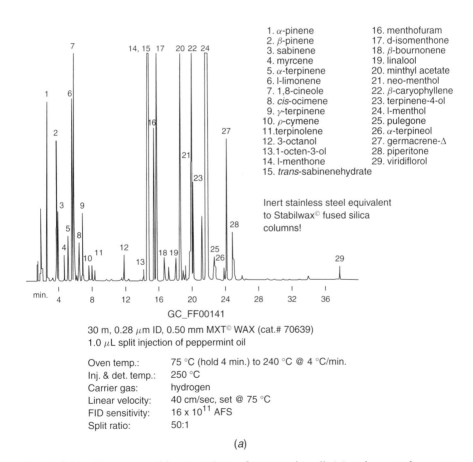

1. α-pinene
2. β-pinene
3. sabinene
4. myrcene
5. α-terpinene
6. l-limonene
7. 1,8-cineole
8. cis-ocimene
9. γ-terpinene
10. p-cymene
11. terpinolene
12. 3-octanol
13. 1-octen-3-ol
14. l-menthone
15. trans-sabinenehydrate
16. menthofuram
17. d-isomenthone
18. β-bournonene
19. linalool
20. minthyl acetate
21. neo-menthol
22. β-caryophyllene
23. terpinene-4-ol
24. l-menthol
25. pulegone
26. α-terpineol
27. germacrene-Δ
28. piperitone
29. viridiflorol

Inert stainless steel equivalent to Stabilwax© fused silica columns!

GC_FF00141

30 m, 0.28 μm ID, 0.50 mm MXT© WAX (cat.# 70639)
1.0 μL split injection of peppermint oil

Oven temp.:   75 °C (hold 4 min.) to 240 °C @ 4 °C/min.
Inj. & det. temp.:   250 °C
Carrier gas: hydrogen
Linear velocity: 40 cm/sec, set @ 75 °C
FID sensitivity: 16×10^{11} AFS
Split ratio: 50:1

(a)

Figure 3.11 Chromatographic separations of peppermint oil (a) and a petroleum wax (b) on MXT columns. (Courtesy of the Restek Corporation.) (*Continued*)

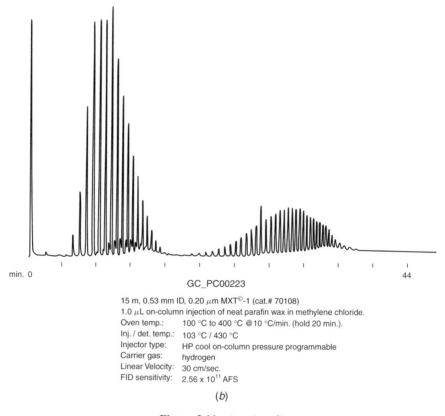

GC_PC00223

15 m, 0.53 mm ID, 0.20 μm MXT©-1 (cat.# 70108)
1.0 μL on-column injection of neat parafin wax in methylene chloride.

Oven temp.:	100 °C to 400 °C @10 °C/min. (hold 20 min.).
Inj. / det. temp.:	103 °C / 430 °C
Injector type:	HP cool on-column pressure programmable
Carrier gas:	hydrogen
Linear Velocity:	30 cm/sec.
FID sensitivity:	2.56 x 10^{11} AFS

(*b*)

Figure 3.11 (*continued*)

When this column is deactivated with a precoating of Carbowax 20M, the residual surface column is reduced considerably (Figure 3.13*b*).

A variety of agents and procedures have been explored for deactivation purposes (26–40). For subsequent coating with nonpolar and moderately polar stationary phases such as polysiloxanes, fused silica has been deactivated by silylation at elevated temperatures, thermal degradation of polysiloxanes and polyethylene glycols, and the dehydrocondensation of silicon hydride polysiloxanes (37, 41–45).

Blomberg and Markides have published a comprehensive review of deactivating methods using polysiloxanes (46). One approach has been suggested by Schomburg et al. (43), who prepared columns having excellent thermal stability with polysiloxane liquid phases as deactivators and proposed that the decomposition products formed at the elevated temperatures chemically bond to surface silanols. Surface–stationary phase compatibility has also been achieved with cyclic siloxanes having the same side functional groups as those of the silicone stationary phase. Octamethylcyclotetrasiloxane (D_4) has been decomposed at 400°C by Stark et al. (47), who postulated that the process involved opening the D_4 ring to form a 1,4-hydroxyoctamethyltetrasiloxane. They indicate that a terminal

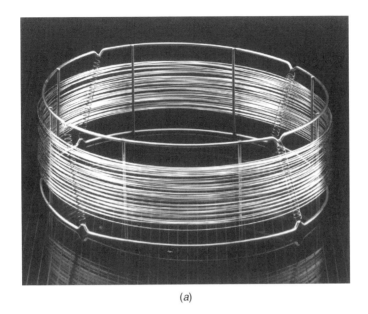

(a)

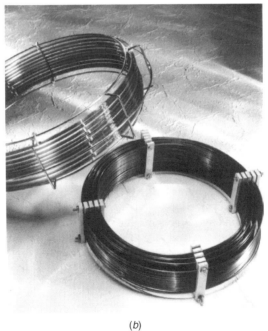

(b)

Figure 3.12   Metal-clad capillary columns:(a) aluminum-clad capillary column; (b) fused silica–lined stainless steel capillary column (lower) and polyimide-clad fused-silica capillary columns (upper). [(a) Courtesy of the Quadrex Corporation; (b) courtesy of the Restek Corporation.]

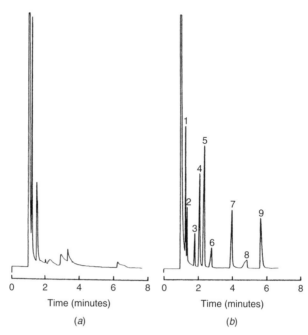

Figure 3.13   Chromatograms of an activity mixture on 15 m × 0.25 mm (*a*) uncoated fused silica and (*b*) fused-silica capillary column after deactivation with Carbowax 20M. Column temperature: 70°C; 25 cm/s He; split injection (100 : 1). Peaks:(1) *n*-dodecane, (2) *n*-tridecane, (3) 5-nonanone, (4) *n*-tetradecane, (5) *n*-pentadecane, (6) 1-octanol, (7) napthalene, (8) 2,6-dimethylaniline, and (9) 2,6-dimethylphenol. (From ref. 104.)

hydroxyl group interacts with a protruding silanol group eliminating water, and in a secondary reaction the other hydroxyl reacts with another silanol or even a tetrasiloxane. Well-deactivated capillary columns can be prepared by this technique (48). Woolley et al. outlined an easily implemented deactivation procedure employing the thermal degradation of polyhydrosiloxane at about 260°C, where the silyl hydride groups undergo reaction with surface silanols to form rather stable Si–O–Si bonds and hydrogen gas (44). This method has the merits of a reaction time below an hour, a relatively low reaction temperature, and a high degree of reproducibility. A representation of selected deactivated surface textures is displayed in Figure 3.14.

Carbowax 20M has also been used successfully to deactivate column surfaces (50,51). After coating a thin film of Carbowax 20M, for example, on the column wall, the column is heated to 280°C, then extracted exhaustively with solvent, leaving a nonextractable film of Carbowax 20M on the surface. Both apolar and polar stationary phases, including Carbowax, can then be coated on capillaries subjected to this pretreatment (52). Dandeneau and Zerenner used this procedure to deactivate their first fused-silica columns (4). Other polyethylene glycols used for deactivating purposes have been Carbowax 400 (53), Carbowax 1000 (54),

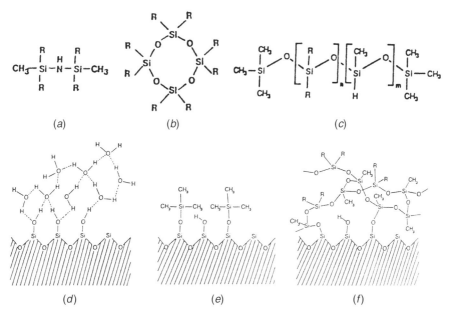

Figure 3.14   Reagents used for deactivation of silanol groups: (*a*) disilazanes, (*b*) cyclic siloxanes (*c*) silicon hydride polysiloxanes. Lower portion is a view of a fused-silica surface with (*d*) adsorbed water (*e*) after deactivation with a trimethylsilylating reagent and (*f*) after treatment with a silicon hydride polysiloxane. (From ref. 49, with permission of Elsevier Science Publishers.)

and Superox-4 (55). Moreover, when a polar polymer is used for deactivation, it may alter the polarity of the stationary phase, and this effect becomes particularly problematic with a thin film of a nonpolar phase where the resulting phase has retentions of a mixed phase. Furthermore, silazanes and cyclic silazanes, as deactivating agents, ultimately yield a basic final column texture, whereas chlorosilanes, alkoxysilanes, hydrosilanes, hydrosiloxanes, siloxanes, and Carbowax produce an acidic column. In essence, a deactivation procedure imparts different residual surface characteristics and is often selected with the stationary phase as well as the application in mind. Many column manufacturers offer base-deactivated columns (and companion base-deactivated inlet liners) with several stationary phases for successful chromatography of amines. There have been two additional approaches to deactivation: (1) the coating of a layer of polypyrrone on the inner surface of the tubing prior to the deposition of the stationary phase, thereby circumventing the temperature limitation of polyimide in high-temperature applications (56), and (2) the Siltek process, in which deactivation is achieved via a vapor deposition process (57,58) as opposed to procedures using, for example, a liquid silazane or chlorosilane. In addition, Siltek deactivation has been utilized for inlet liners, borosilicate glass wool, and guard columns, discussed later in the chapter. An alternative procedure is the utilization of OH-terminated stationary phases where deactivation and immobilization of the phase occurs in a single-step process. Still another approach, which perhaps offers great overall advantages, is the sol-gel method, where the primary advantage is that the column is made in a single process rather than in a series of steps. The single step incorporates column deactivation, coating of the stationary phase, and immobilization of the coated film. Sol-gel columns are discussed further later in the chapter.

Static Coating of Capillary Columns

The goal in coating a capillary column is the uniform deposition of a thin film, typically ranging from 0.1 to 8 μm in thickness, on the inner wall of a length of clean deactivated fused-silica tubing. Jennings (59) has reviewed the various methods for coating stationary phases. The static method of coating is discussed here because it is most widely used today by column manufacturers.

In this procedure, first described by Bouche and Verzele (60), the column is completely filled initially with a solution of known concentration of stationary phase. One end of the column is sealed and the other is attached to a vacuum source. As the solvent evaporates, a uniform film is deposited on the column wall. The column must be maintained at constant temperature for uniform film deposition. The coating solution should be free of microparticulates and dust, be degassed so that bumping does not occur during solvent evaporation, and there should be no bubbles in the column. Pentane is the solvent recommended, because of its high volatility, and should be used wherever stationary-phase solubility permits. Evaporation time is approximately half of that required to evaporate methylene chloride. The static coating technique offers the advantage of an accurate determination of the phase ratio (Section 3.4), from which the film thickness of the stationary phase can be calculated.

Capillary Cages

Since the ends of flexible fused-silica capillary tubing are inherently straight, columns must be coiled and confined on a circular frame, also called a *cage* (Figure 1.2*C*). The capillary column can then be mounted securely in the column oven of a gas chromatograph. Fused-silica capillary columns 0.10 to 0.32 mm in i.d. are wound around a 5- or 7-in.-diameter cage, whereas an 8-in. cage is used with megabore columns (0.53 mm i.d.). Four-inch-diameter cages are available for specific instrumentation and applications; however, a megabore capillary cannot be wound on a cage of this diameter. Installation of a capillary column is greatly facilitated, since the ends of a fused-silica column can easily be inserted into sample inlets and detector systems at the appropriate lengths recommended. The ultimate in gas chromatographic system inertness is attainable with on-column injection, where a sample encounters only fused silica from the point of injection to the tip of an FID flame jet.

Test Mixtures for Monitoring Column Performance

The performance of a capillary column can be evaluated with a test mixture whose components and resulting peak shapes serve as monitors of column efficiency and diagnostic probes for adverse adsorptive effects and the acid–base character of a column. These mixtures are used by column manufacturers in the quality control of their columns and are recommended for the chromatographic laboratory.

A chromatogram of a test mix and a report are usually supplied with a commercially prepared column. Using the chromatographic conditions indicated, the separation should be duplicated by the user prior to running samples with column. In the test report evaluating the performance of the column, chromatographic data are listed. These may include retention times of the components in the text mix; corresponding Kovatś retention index values (Chapter 2) of several, if not all, of the solutes; the number of theoretical plates N and/or the effective plate number N_{eff}; and the acid–base inertness ratio (the peak-height ratio of the acidic and basic probes in the test mixture). The values of two additional chromatographic parameters, separation number (Trennzahl number) and coating efficiency, may also be included in the report; the significance of TZ was discussed in Chapter 2, and coating efficiency is treated in the next section. Commonly used components in a stringent test mixture, their accepted abbreviations, and functions are listed in Table 3.7. A selection of commercially available general test mixtures and their designed general purposes appear in Table 3.8.

TABLE 3.7   Test Mixture Components and Role

Probe[a]	Function
n-Alkanes, typically C10–C15	Column efficiency; Trennzahl number
Methyl esters of fatty acids, usually C9–C12 (E9–E12)	Separation number; column efficiency
1-Octanol (ol)	Detection of hydrogen-bonding sites, silanol groups
2,3-Butanediol (D)	More rigorous test of silanol detection
2-Octanone	Detection of activity associated with Lewis acids
Nonanal (al)	Aldehyde adsorption other than via hydrogen bonding
2,6-Dimethylphenol (P)	Acid–base character
2,6-Dimethylaniline (A)	Acid–base character
4-Chlorophenol	Acid–base character
n-Decylamine	Acid–base character
2-Ethylhexanoic acid (S)	More stringent measure of irreversible adsorption
Dicyclohexylamine (am)	More stringent measure of irreversible adsorption

[a] Abbreviations of the components in the comprehensive Grob mix are indicated in parentheses; see Table 3.8.

TABLE 3.8 Test Mixtures Used to Evaluate GC Columns

Grob Test Mix for use under temperature programming conditions (Restek); similar mixture available from Supelco

n-C10-FAME	0.42 mg/mL	2,6-dimethylphenol	0.32 mg/mL
n-C11-FAME	0.42 mg/mL	2-ethylhexanoic acid	0.38 mg/mL
n-C12-FAME	0.41 mg/mL	nonanal	0.40 mg/mL
2,3-butanediol	0.53 mg/mL	1-octanol	0.36 mg/mL
dicyclohexylamine	0.31 mg/mL	n-undecane	0.29 mg/mL
2,6-dimethylaniline	0.32 mg/mL	n-decane	0.28 mg/mL

Nonpolar Isothermal Column Test Mixture (Supelco); 0.50 mg/mL of each component in methylene chloride; for all nonpolar phases

2-octanone	1-octanol
n-undecane	n-dodecane
n-decane	n-tridecane
2,6-dimethylaniline	2,6-dimethylphenol

Intermediate Polarity Isothermal Test Mixture (Supelco); 0.50 mg/mL of each component in methylene chloride; for all intermediate-polarity polysiloxane phases containing greater than 20% phenyl content

n-decane	n-tridecane
n-dodecane	1-octanol
2-octanone	2,6-dimethylaniline
2,6-dimethylphenol	n-tetradecane
n-undecane	

Polar Column Isothermal Test Mixture (Supelco); 0.50 mg/mL of each component in methylene chloride

2-octanone	2,6-dimethylphenol
n-octadecane	n-hexadecane
n-pentadecane	n-eicosane
2,6-dimethylaniline	n-heptadecane
1-octanol	

General Isothermal Test Mixture (Restek)

1,2-hexanediol	0.46 mg/mL	1-octanol	0.36 mg/mL
n-decane	0.29 mg/mL	nonanal	0.40 mg/mL
n-undecane	0.29 mg/mL	2,6-dimethylaniline	0.32 mg/mL
n-dodecane	0.29 mg/mL	2,6-dichlorophenol	0.57 mg/mL
n-tridecane	0.29 mg/mL	naphthalene	0.32 mg/mL

Amine Column Test Mixture (Restek)

1,2-butanediol	0.60 mg/mL	diethanolamine	1.20 mg/mL
pyridine	0.60 mg/mL	2-nonanol	0.60 mg/mL
n-decane	0.60 mg/mL	2,6-dimethylaniline	0.60 mg/mL
diethylenetriamine	1.20 mg/mL	n-dodecane	0.60 mg/mL

Amine Column Test Mixture (Supelco); 0.50 mg/mL of each component in methyl t-butyl ether

2,4-dimethylaniline	n-heptadecane
2,6-dimethylaniline	n-hexadecane
tri-n-hexylamine	n-nonylamine
n-benzylamine	n-octadecane
n-decylamine	n-octylamine
n-eicosane	n-pentadecane

TABLE 3.8   (*continued*)

Fragrance Materials Text Mixture (Restek); proposed by Fragrance Materials Association

benzyl salicyclate	362 parts	geraniol	6 parts
cinnamic aldehyde	5 parts	hydroxycitronella	50
cinnamic alcohol	3 parts	*d*-limonene	200
ethyl butyrate	3 parts	thymol crystal	3
eucalyptol	5 parts	vanillin	1
cinnamic acetate	3 parts	benzoic acid	1% of mix

Test Mixture for α-Cyclodextrin Chiral Columns (Supelco); 0.50 mg/mL of each
   component in methylene chloride

(-)-1,2-propanediol	*n*-undecane
(+)-1,2-propanediol	*m*-xylene
n-decane	*p*-xylene
n-nonane	

Test Mixture for β-Cyclodextrin Chiral Columns (Supelco); 0.50 mg/mL of each
   component in methylene chloride

(+)-3,3-dimethyl-2-butanol	*n*-decane
(+)-3-methyl-2-heptanone	*n*-nonane
1-hexanol	*n*-undecane

Diagnostic Role Played by Components of Test Mixtures

The first chromatogram obtained on a new column may be viewed as the "birth certificate" of a column and defines column performance at time $t = 0$ in the laboratory; a test mix should also be analyzed periodically to determine any changes in column behavior that may be occurring with age and use. For example, a column may acquire a pronouncedly basic character if it has been employed routinely for amine analyses. Another important but often overlooked aspect is that a test mixture serves to monitor the performance of the *total* chromatographic system, not just the performance of the column. If separations gradually deteriorate over time, the problem may not always be column-related but could be due to extracolumn effects, such as a contaminated or activated inlet liner or an incorrect column reinstallation procedure. An ideal capillary column should be well deactivated, with excellent thermal stability and high separation efficiency. The extent of deactivation is usually manifested by the amount of peak tailing for polar compounds. The most comprehensive and exacting test mixture is the solution reported by Grob et al. (61) and is more sensitive to residual surface activity than other polarity mixes. Adsorption may cause (1) broadened peaks of Gaussian shapes, (2) a tailing peak of more or less the correct peak area, (3) a reasonably shaped peak with reduced area, and (4) a skewed peak of correct area but having an increased retention time. Furthermore, irreversible adsorption cannot always be detected by peak shape. In the Grob procedure one measures peak heights as a percentage of that expected for complete and undistorted elution. The technique encompasses all types of peak deformations (broadening, tailing,

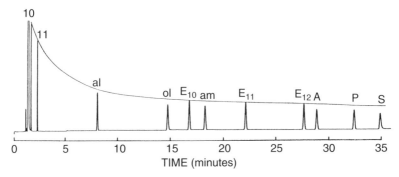

Figure 3.15   Chromatogram of a comprehensive Grob mixture on a 15 m × 0.32 mm i.d. Carbowax 20M capillary column. Column conditions: 75 to 150°C at 1.7°C/min; 28 cm/s He. Designation of solutes appears in Table 3.7. (From ref. 104.)

and irreversible adsorption). A solution whose components are present at specific concentrations is analyzed under recommended column temperature-programming conditions.

In practice, the percentage of the peak height is determined by drawing a line (the 100% line) connecting the peak maxima of the nonadsorbing peaks (*n*-alkanes and methyl esters), as shown in Figure 3.15. Alcohols are more sensitive than the other probes to adsorption caused by hydrogen bonding to exposed silanols. The acid and base properties are ascertained with probe solutes such as 2,6-dimethylaniline and 2,6-dimethylphenol, respectively. However, most column manufacturers recommend a modification of the Grob scheme to circumvent the lengthy time involved and, instead, tailor the composition of the mix and column temperature conditions to be commensurate with the particular deactivation procedure and stationary phase under consideration. In addition to the mixture components listed in Table 3.8, a widely used test mixture consists of the components designated in Figure 3.16, where the test mix is also used here to demonstrate selectivity by comparing separations on three columns of the same dimensions that have different stationary phases.

The effects of deactivation, the chemistry of the stationary phase and its cross-linking, as well as the effect of any postprocess treatment all appear in the final version of a column (62). An example of this situation is separation of a Grob mixture (Figure 3.17*a*) performed on a 15-m × 0.25-mm-i.d. fused-silica capillary column deactivated with Carbowax 20M, after which the column received a recoat of the polymer. After cross-linking of the stationary phase (Figure 3.17*b*) the column behavior changed markedly. The 2,3-butanediol peak (D), absent in Figure 3.17*a*, is present on the cross-linked phase, which has acquired increased acidity in the cross-linking process. Note the decreased peak height of the dicyclohexylamine probe (am) and the increased peak height of 2-ethylhexanoic acid (S). Thus, any change or a minor modification in column preparation can affect the final column performance.

1) 2-octanone
2) n-decane
3) 1-octanol
4) 2,6-DMP
5) n-undecane
6) 2,6-DMA
7) naphthalene
8) n-dodecane
9) n-tridecane

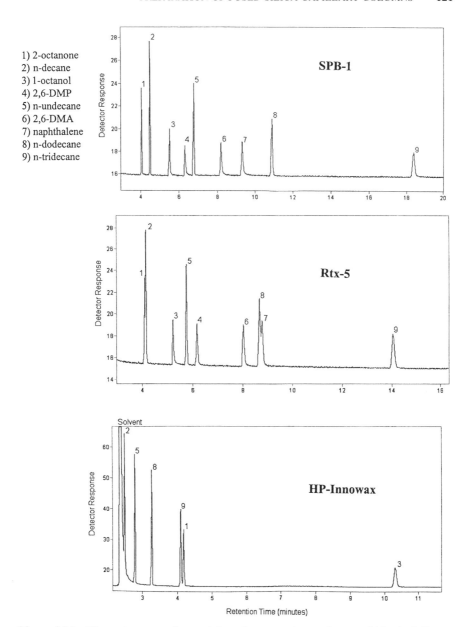

Figure 3.16 Chromatograms of an activity mixture on three columns of identical dimensions but different stationary phases, as indicated; conditions: 15 m × 0.25 mm i.d. × 0.25 − μm film capillary columns, 110°C, 25 cm/s He, FID.

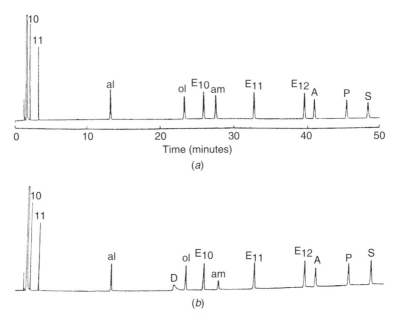

Figure 3.17   Chromatogram of a comprehensive Grob mixture on a 15 m × 0.32 mm i.d. Carbowax 20M capillary column (*a*) after coating and (*b*) after cross-linking the stationary phase. Column conditions: 75 to 150°C at 2°C/min; 28 cm/s He. Designation of solutes appears in Table 3.16. (From ref. 104.)

3.4   CHROMATOGRAPHIC PERFORMANCE OF CAPILLARY COLUMNS

Golay Equation Versus the van Deemter Expression

The fundamental equation underlying the performance of a gas chromatographic column is the van Deemter expression [equation (2.1)], which may be expressed as

$$H = A + \frac{B}{\bar{u}} + C\bar{u} \tag{3.3}$$

where H is the height equivalent to a theoretical plate, A the eddy diffusion or multiple-path term, B the longitudinal diffusion contribution, $\bar{u}$ the average linear velocity of the carrier gas, and C the resistance to mass transfer term. In the case of a capillary column, the A term is equal to zero because there is no packing material. Thus, equation 3.3 simplifies to

$$H = \frac{B}{\bar{u}} + C\bar{u} \tag{3.4}$$

This abbreviated expression is often referred to as the *Golay equation* (43). The B term may be expressed as $2D_g\bar{u}$, where D_g is the binary diffusion coefficient of the solute in the carrier gas. Peak broadening due to longitudinal diffusion is a consequence of the residence time of the solute within the column and the nature of the carrier gas. This effect becomes pertinent only at low linear velocities or flow rates and is less pronounced at high velocities (Figure 2.11).

However, the major contributing factor contributing to band broadening is the C term, in which the resistance to mass transfer can be represented as the composite of the resistance to mass transfer in the mobile phase C_g and that in the stationary phase C_l:

$$C = C_g + C_l \tag{3.5}$$

where

$$C_g = \frac{r^2(1 + 6k + 11k^2)}{D_g 24(1 + k)^2} \tag{3.6}$$

$$C_l = \frac{2kd_f^2 \bar{u}}{3(1 + k)^2 D_l} \tag{3.7}$$

D_l is the diffusion coefficient of the solute in the stationary phase, k the retention factor of the solute, d_f the film thickness of stationary phase, and r the radius of the capillary column. With capillary columns, C_l is small and becomes significant only with capillary columns having a thick film of stationary phase. The Golay equation may then be rewritten as

$$H = \frac{B}{\bar{u}} + C_g \bar{u} = \frac{2D_g}{\bar{u}} + \frac{r^2(1 + 6k + 11k^2)\bar{u}}{D_g 24(1 + k)^2} \tag{3.8}$$

The optimum linear velocity corresponding to the minimum in a plot of H versus u (Figure 2.20) can be obtained by setting $dH/du = 0$ and solving for $\bar{u}$:

$$\frac{dH}{du} = 0 = \frac{B}{\bar{u}^2} + C_g \tag{3.9}$$

Thus, $u_{\text{opt}} = (B/C_g)^{1/2}$ and the value of H corresponding to this optimum linear velocity, H_{min}, is

$$H_{\text{min}} = r\sqrt{\frac{1 + 6k + 11k^2}{3(1 + k)^2}} \tag{3.10}$$

Consequently, as the diameter of a capillary column decreases, both maximum column efficiency N and maximum effective efficiency N_{eff} increase and are dependent on the particular solute retention k. Retention in capillary GC is usually expressed as the retention factor, k, where

$$k = t_R - t_M/t_M$$

TABLE 3.9 Column Efficiency as a Function of Inner Diameter and Retention Factor

Inner Diameter (mm)	k	h_{min}	Maximum Plates per Meter, N	Effective Plates per Meter, N_{eff}
0.10	1	0.061	16,393	4,098
	2	0.073	13,697	6,027
	5	0.084	11,905	6,667
	10	0.090	11,111	9,222
	20	0.093	10,752	9,784
	50	0.095	10,526	10,105
0.25	1	0.153	6,536	1,634
	2	0.182	5,495	2,442
	5	0.210	4,762	3,307
	10	0.224	4,464	3,689
	20	0.231	4,329	3,925
	50	0.236	4,237	4,073
0.32	1	0.196	5,102	1,276
	2	0.232	4,310	1,896
	5	0.269	3,717	2,082
	10	0.286	3,497	2,903
	20	0.296	3,378	3,074
	50	0.302	3,311	3,179
0.53	1	0.325	3,076	769
	2	0.384	2,604	1,146
	5	0.445	2,247	1,258
	10	0.474	2,110	1,751
	20	0.490	2,041	1,857
	50	0.500	2,000	1,920

In Table 3.9 the effect of column inner diameter on maximum attainable column efficiency N is presented as a function of the retention factor. For a capillary of a given inner diameter, one can see that there is an increase in plate height with increasing k, with a corresponding decrease in plate number and an increase in effective plate count. As the inner diameter of a capillary column increases, column efficiency N drops markedly while the effective plate number N_{eff} increases. For separations requiring high resolution, columns of small inner diameter are recommended. Expressing efficiency in terms of plates per meter allows the efficiency of columns of unequal lengths to be compared. Also included in Table 3.9 are data for 0.53 mm i.d., the diameter of the megabore capillary column, which has been termed an alternative to the packed column. The merits and features of this particular type of column are discussed in Section 3.7.

Choice of Carrier Gas

Capillary column efficiency is dependent on the carrier gas used, the length and inner diameter of the column, the retention factor of the particular solute selected for calculation of the number of theoretical plates, and the film thickness of the

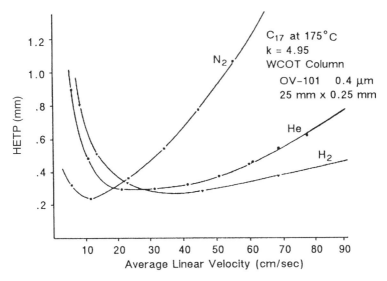

Figure 3.18   Profiles of HETP versus linear velocity for the carrier gases: helium, hydrogen, and nitrogen. (Courtesy of Agilent Technologies.)

stationary phase. Profiles of H versus u for three carrier gases with a thin-film capillary column are displayed in Figure 3.18. Although the lowest minimum, and therefore the greatest efficiency, are obtained with nitrogen, speed of analysis must be sacrificed, as shown in Figure 3.19. The increasing portion of the curve is steeper for nitrogen in Figure 3.18, which necessitates working at or near u_{opt}; otherwise, loss in efficiency (and resolution) quickly results. On the other hand, if one is willing to accept a slight loss in the number of theoretical plates, a more favorable analysis time is possible with helium and hydrogen as carrier gases, because u_{opt} occurs at a higher linear velocity. Moreover, the mass transfer contribution or rising portion of a curve is less steep with helium or hydrogen, which permits working over a wider range of linear velocities without substantial sacrifice in resolution. This advantage becomes evident in comparing the capillary separation of the components in calmus oil with nitrogen and hydrogen as carrier gases in Figure 3.20.

In comparing these carrier gases, another benefit becomes apparent at linear velocities corresponding to equal values of plate height. With the lighter carrier gases, solutes can be eluted at lower column temperatures during temperature programming with narrower band profiles, since higher linear velocities can be used. Thus, either helium or hydrogen is recommended over nitrogen, and indeed these gases are used today as carrier gases for capillary gas chromatography. One advantage of using hydrogen is that the plate number varies less for hydrogen than for helium as linear velocity increases. The use of hydrogen for any application in the laboratory always requires safety precautions in the event of a leak. Precautionary measures should be taken for the safe discharge of hydrogen from the split vent in the split-injection mode.

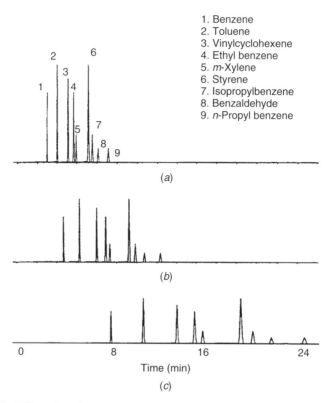

1. Benzene
2. Toluene
3. Vinylcyclohexene
4. Ethyl benzene
5. *m*-Xylene
6. Styrene
7. Isopropylbenzene
8. Benzaldehyde
9. *n*-Propyl benzene

Figure 3.19   Effect of carrier gas on separation at optimum linear velocities. (From ref. 63, reprinted with permission of John Wiley & Sons, Inc.)

Measurement of Linear Velocity and Flow Rate

The flow rate through a capillary column whose inner diameter is less than 0.53 mm is difficult to measure accurately and reproducibly by a conventional soap-bubble meter. Instead, the flow of carrier gas through a capillary column is usually expressed as a linear velocity rather than as a volumetric flow rate. Linear velocity may be calculated by injecting a volatile nonretained solute and noting its retention time, t_M (seconds). For a capillary column of length L in centimeters,

$$u(cm/sec) = \frac{L}{t_M} \qquad (3.11)$$

For example, the linear velocity of carrier gas through a 30-m column where methane has a retention time of 2 minutes is 3000 cm per 120 s or 25 cm/s. If desired, the volumetric flow rate F (mL/min) may be computed from the relationship

$$F(\text{mL/min}) = 60\pi r^2 u \qquad (3.12)$$

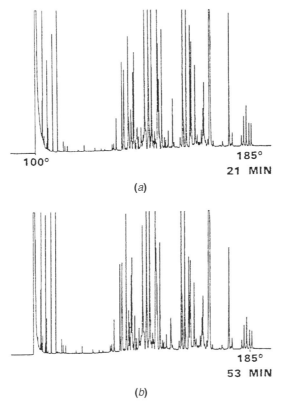

Figure 3.20 Chromatograms of the separation of calmus oil using (*a*) hydrogen as carrier gas, 4.2 mL/min at a programming rate of 4.0°C/min and (*b*) nitrogen as carrier gas, 2.0 mL/min programming rate of 1.6°C/min on a 40 m × 0.3 mm i.d. capillary column (0.12-μm film). (From ref. 64, with permission of Alfred Heuthig Publishers.)

where r is the radius of the column in centimeters. An injection of methane is convenient to use with a FID to determine t_M, or a headspace injection of methylene chloride and acetonitrile can be made with an ECD and NPD, respectively. Nitrogen and oxygen (air) may be used with a mass spectrometer while ethylene or acetylene vapors can be injected with a photoionization detector (PID). Linear velocities and flow rates of helium and hydrogen recommended for capillary columns of various inner diameters are listed in Table 3.10.

Effect of Carrier Gas Viscosity on Linear Velocity

Chromatographic separations using capillary columns are achieved under constant-pressure conditions, as opposed to packed columns, which are usually operated in a flow-controlled mode. The magnitude of the pressure drop across a capillary column necessary to produce a given linear velocity is a function of the particular carrier

TABLE 3.10   Recommended Linear Velocities and Flow Rates with Helium and Hydrogen[a]

Inner Diameter (mm)	Linear Velocity (cm/s)		Flow Rate (mL/min)	
	Helium	Hydrogen	Helium	Hydrogen
0.18	20–45	40–60	0.3–0.7	0.6–0.9
0.25	20–45	40–60	0.7–1.3	1.2–2.0
0.32	20–45	40–60	1.2–2.2	2.2–3.0
0.53	20–45	40–60	4.0–8.0	6.0–9.0

Source: Data from the 1994–1995 J&W Scientific Catalog.
[a]Column length 30 m.

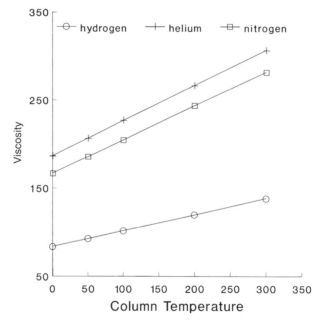

Figure 3.21   Effect of temperature on carrier gas viscosity. (Data for curves generated from viscosity–temperature relationships in ref. 49.)

gas and length to inner diameter of the column. The relationship between viscosity and temperature for any gas is linear, as shown in Figure 3.21 for helium, hydrogen, and nitrogen. In gas chromatography, as the column temperature increases, the linear velocity decreases because of the increased viscosity of the carrier gas. Thus, initially, higher linear velocities are established for temperature-programmed analyses than for isothermal separations. If we compare columns of identical dimensions and operate them at the same inlet pressure and temperature, the linear velocity will be highest for hydrogen and lowest for helium. Therefore, whenever a change in the type of carrier gas is made in the laboratory, linear velocities (or flow rates for packed columns) should actually be measured and one should not reconnect the pressure regulator using the same delivery pressure.

Phase Ratio

In addition to the nature of the carrier gas, column efficiency and ultimately, the resolution and sample capacity of a capillary column are affected by the physical nature of the column: namely, the inner diameter and film thickness of the stationary phase. An examination of the distribution coefficient K_D as a function of chromatographic parameters is helpful here. K_D is constant for a given solute–stationary phase pair and is dependent only on column temperature. K_D may be defined as

$$K_D = \frac{\text{concentration of solute in stationary phase}}{\text{concentration of solute in carrier gas}} \qquad (3.13)$$

or

$$K_D = \frac{\text{amount of solute in stationary phase}}{\text{amount of solute in mobile phase}}$$
$$\times \frac{\text{volume of carrier gas}}{\text{volume of stationary phase in the column}} \qquad (3.14)$$

K_D can now be expressed as

$$K_D = k\beta = \frac{\pi}{2d_f} \qquad (3.15)$$

where β is the phase ratio and is equal to $r/2d_f$, r is the radius of the column, and d_f is the film thickness of stationary phase. At a given column temperature, retention increases as the phase ratio of the column decreases, which can be manipulated either by decreasing the diameter of the column or by increasing the film thickness of the stationary phase; similarly, a decrease in retention is noted with an increase in β. Since K_D is a constant at a given column temperature, film thickness and column diameter play key roles in determining separation power and sample capacity. In selecting a capillary column, the phase ratio should be considered.

As the film thickness decreases, k, the retention factor, also decreases at constant temperature, column length, and i.d. Conversely, with an increase in film thickness in a series of columns having the same dimensions, retention increases under the same temperature conditions. This effect of film thickness on separation is demonstrated in the series of parallel chromatograms appearing in Figure 3.22. Column diameter limits the maximum amount of stationary phase that can be coated on its inner wall. Small-diameter columns usually contain thinner films of stationary phase, while thicker films can be coated on wider-bore columns. The concept of phase ratio allows two columns of equal length to be compared in terms of sample

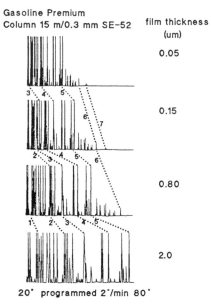

Gasoline Premium
Column 15 m/0.3 mm SE-52

Figure 3.22 Chromatograms of gasoline on capillary columns with varying film thickness of stationary phase, SE-52. (From ref. 64, with permission of Alfred Heuthig Publishers.)

TABLE 3.11   Selected Phase Ratios as a Function of Column Inner Diameter and Film Thickness

Film Thickness (μm)	Column i.d. (mm)					
	0.10	0.18	0.25	0.32	0.45	0.53
0.10	250	450	625	800	1130	1325
0.25	100	180	250	320	450	530
0.50	50	90	125	160	225	265
1.0	25	45	63	80	112	128
3.0	17	30	21	27	38	43
5.0	5	9	13	16	23	27

capacity and resolution. Phase ratios of columns having different inner diameters and film thicknesses are listed in Table 3.11.

As depicted in Table 3.9, column efficiency increases as column diameter decreases. Sharper peaks yield improved detection limits. However, as column diameter decreases, so does sample capacity, as indicated in Table 3.12. Column temperature conditions and the linear velocity of the carrier gas can usually be adjusted to have a more favorable time of analysis. In Figure 3.23 these parameters are placed into perspective in a pyramidal format as a function of the inner diameter of a capillary column.

TABLE 3.12   Column Capacity as a Function of Inner Diameter and Film Thickness

Inner Diameter (mm)	Film Thickness (μm)	Capacity[a] (ng/component)
0.25	0.15	60–70
	0.25	100–150
	0.50	200–250
	1.0	350–400
0.32	0.25	150–200
	0.5	250–300
	1.0	400–450
	3.0	1200–1500
0.53	1.0	1000–1200
	1.5	1400–1600
	3.0	3000–3500
	5.0	5000–6000

Source: Data from the 1994–1995 J&W Scientific Catalog.

[a]Capacity is defined as the amount of component where peak asymmetry occurs at 10% at half-height.

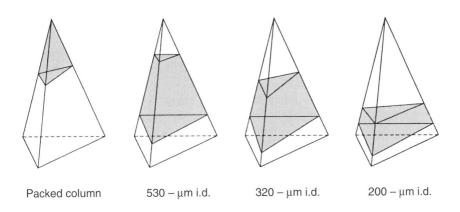

Packed column   530 – μm i.d.   320 – μm i.d.   200 – μm i.d.

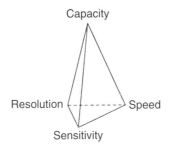

Figure 3.23   Chromatographic pyramids for packed and capillary columns of varying inner diameter. (Courtesy of Agilent Technologies.)

Coating Efficiency

The coating efficiency, also called the *utilization of theoretical efficiency* (UTE), is the ratio of the actual efficiency of a capillary column to its theoretical maximum possible efficiency. Coating efficiency or UTE is expressed as

$$\%\text{Coating efficiency} = \frac{H_{min}}{H} \times 100 \tag{3.16}$$

where H_{min} is as defined in equation 3.11. Coating efficiency is a measure of how well a column is coated with stationary phase. Coating efficiencies of nonpolar columns range from 90 to 100%. Polar columns have somewhat lower coating efficiencies, 60 to 80%, because a polar stationary phase is more difficult to coat uniformly. Also, larger-diameter columns tend to have higher coating efficiencies. This parameter is typically listed on a test report shipped with a new column and is important to column manufacturers for monitoring the quality of their product. It is usually of no concern to most capillary column users.

3.5 STATIONARY-PHASE SELECTION FOR CAPILLARY GAS CHROMATOGRAPHY

Requirements

The use of packed columns for gas chromatographic separations requires having an assortment of columns available with different stationary phases to compensate for column inefficiency by a commensurate gain in selectivity. With a capillary column, the demands on selectivity, although important, are not as stringent, because of the high plate count possible with a capillary column. However, with the transition from the era of the packed column to that of the capillary column, a gradual redefinement in the requirements of the stationary phase took place.

Many liquid phases for packed column purposes were unacceptable for capillary GC. Although they offered selectivity, overriding factors responsible for their disfavor were overall lack of thermal stability and the instability of the stationary phase as a thin film at elevated temperatures and during temperature programming. In the latter processes, it is crucial that the phase remain a thin uniform film; otherwise, loss of both inertness and column efficiency results. Today, these problems have been solved and the refinements are reflected in the high performance of commercial columns. The impetus has been driven by the improvements in the sensitivity of mass spectrometers such that the MS detector is now the second most popular detector in GC (the FID is the most widely used detector). This rise in the use of GC-MS has also necessitated more thermally stable columns offering much less column bleed.

History

The coating of a glass capillary column was achieved by roughening its inner surface prior to coating for enhanced wettability by stationary phases having a wide range of polarities and viscosities, but this option is unavailable with fused silica. The wettability of fused silica proved to be more challenging because its thin wall does not permit aggressive surface modification. Consequently, fewer phases initially could be coated on fused silica than on glass capillaries. Although polar phases could be deposited successfully on glass capillaries, fused-silica columns coated with polar phases were especially inferior in terms of efficiency and thermal stability.

Viscosity of the film of stationary phase after deposition under the thermal conditions of GC proved to be an important consideration. Wright and co-workers (65) correlated viscosity of a stationary phase with coating efficiency and the stability of the coated phase. The results of their study supported the experimental success of viscous gum phases, which yielded higher coating efficiencies and had greater thermal stability than those of their corresponding nonviscous counterparts. The popularity of the nonpolar polysiloxane phases is in part due to the fact that their viscosity is nearly independent of temperature (66). However, the introduction of phenyl and more polar functionalities on the polysiloxane backbone causes a decrease in viscosity of a polysiloxane at elevated temperatures, resulting in thermal instability. Five areas have enhanced the quality of stationary phases for the modern practice of capillary GC:

1. In situ free-radical cross-linking of stationary phases coated on fused silica
2. The synthesis, processing, or commercial availability of a wide array of highly viscous gum phases
3. The use of OH-terminated polysiloxanes
4. The use of polysilphenylene–siloxanes, sometimes referred to as polyarylene–siloxane phases
5. The utilization of sol-gel chemistry

Let us focus now on selection of the stationary phase, the most important aspect in column selection. Factors influencing the choice of inner diameter, film thickness of the stationary phase, and column length are described later in the chapter. In choosing a stationary phase for capillary separations, remember the adage "like dissolves like." As a starting point, try to match the functional groups present in the solutes under consideration with those in a stationary phase, as is the case in the selection of a packed column, of course. In the analysis of polar species, for instance, select a polar stationary phase. Fine-tune this choice, if necessary, by examining McReynolds constants of specific interactions of a particular solute with a stationary phase. However, for reasons elucidated throughout the remainder of this section, a polar phase tends to exhibit slightly less column efficiency than that of a nonpolar phase, has a lower maximum temperature limit, and will have a shorter lifetime if operated for a prolonged period of time at an elevated temperature. The effect of the lower thermal stability of polar phases can be alleviated by selecting a thinner film of stationary phase and a shorter column length for more favorable elution temperatures. Here are some guidelines:

1. Use the least polar phase that will generate the separation needed.
2. Nonpolar phases are in general more "forgiving" than polar phases because they are more resistant to traces of oxygen and water in the carrier gas and basically, oxidation and hydrolysis.
3. Nonpolar phases are more inert than polar phases, bleed less, have a wide range of operating column temperatures, and have higher coating efficiencies.
4. With nonpolar stationary phases, for the most part, separations occur on the basis of boiling points. Introducing or increasing trifluoropropyl, cyanopropyl, or phenyl content, separations become the result of interactions between functional groups, dipoles, charge distributions, and so on.
5. Separations of compounds that differ in their capacity for hydrogen bonding, such as alcohols and aldehydes, can probably best be achieved with polyethylene-type stationary phases.

Additional helpful guidance is provided in Appendix B.

Comparison of Columns from Manufacturers

An extensive list of columns containing different stationary phases commercially available for capillary separations appears in Table 3.13. Included in this handy compilation of columns are commonly used polysiloxanes and polyethylene glycol phases, which are suitable for most applications. The majority of analyses can be performed on columns containing 100% dimethylpolysiloxane or 5% phenyl- and 95% methylpolysiloxane, a cyanopolysiloxane, and a polyethylene glycol. Additional selectivity in a separation can always be achieved by using a trifluoropropylpolysiloxane or phases of varying cyano and phenyl content. Separation of permanent gases and light hydrocarbons can now be performed on a capillary column containing an adsorbent (a porous polymer, alumina, molecular sieves), which serves as a direct substitute of the packed column version and is treated in Section 3.6. Table 3.13 should serve as a handy reference for stationary phases and their chemical composition when comparing columns and chromatographic methods, and the companion Table categorizes columns by application.

Column manufacturers continue to add MS grade, polyarylene–and polyphenylene–siloxane and method-specific columns to their product line on a regular basis; the reader is urged to consult technical information available from vendors or on their Web sites. Perusal of Tables 3.13 and indicates several trends in the bewildering array of column designations, which in itself is testimony to the widespread use of capillary GC. Each manufacturer has its own alphanumeric designation for is product line [e.g., HP (HP-), J&W (DB-), Restek (RT-), SGE (BD-), Supelco (SPB-)]. In many instances, the numerical suffix corresponds to the numerical suffix of the appropriate OV (Ohio Valley) and is representative of the percentage of the polar functional group or modifier in a given polysiloxane. For example, Rtx-5 is listed as being chemically similar to BP-5 and DB-5 in terms of its chromatographic properties, including its selectivity and retention characteristics. For a given

TABLE 3.13 Cross-Reference of Capillary Columns Offered by Manufacturers

Restek	Phase Composition	Agilent/ J&W	Supelco	Alltech	SGE	Varian/ Chrompack	Phenomenex	Quadrex	Perkin-Elmer	USP Nomenclature[a]
Rtx-1, Rtx-1 MS	100% dimethyl-polysiloxane	HP-1, HP-101, DB-1, DB-1ht, SE-30	Equity-1, SPB-1, SP-2100, SPB-1 Sulfur, SE-30	AT-1, SE-30, AT-1 MS	BP-1	VF-1 MS, CP Sil 5 CB, CP Sil 5 CB MS	ZB-1	007-1	PE-1	G1, G2, G38
Rtx-5, XTI-5, Rtx-5 MS	95% Dimethyl- and 5% diphenylpoly-siloxane	HP-5, PAS-5, DB-5, DB-5.625, DB-5ht, SE-54	Equity-5, SPB-5, PTE-5, SE-54, SAC-5, PTE-5 QTM	AT-5, SE-54, AT-5 MS	BP-5	VF-5 MS, CP Sil 8 CB, CP Sil 8 CB MS	ZB-5	007-2	PE-2	G27, G36
Rtx-5Sil MS	95% Dimethyl- and 5% diphenylpoly-silarylene	HP-5TA, DB-5 ms	MDN-5S		BPX-5	VF-XMS	ZB-5 MS			
Rtx-1301, Rtx-624	6% Cyanopropyl-phenyl- and 94% dimethylpoly-siloxane	HP-1301, HP-624, DB-1301, DB-624	SPB-1301	AT-624	BP-624		ZB-624	007-1301		G43
Rtx-20	80% Dimethyl- and 20% diphenyl-polysiloxane		SPB-20, VOCOL	AT-20				007-7	PE-7	G28, G32
Rtx-35, Rtx-35 MS	65% Dimethyl- and 35% diphenyl-polysiloxane	HP-35, DB-35	SPB-35, SPB-608	AT-35, AT-35 MS	BPX-35, BPX-608	VF-35 MS	ZB-35	007-11	PE-11	G42
Rtx-1701	14% Cyanopropy-lphenyl 86% dimethylpoly-siloxane	HP-1701, PAS-1701, DB-1701	SPB-1701	AT-1701	BP-10	CP Sil 19 CB	ZB-1701	007-1701	PE-1701	G46
Rtx-200, Rtx-200 MS	Trifluoropropyl-methylpolysilox-ane	DB-210, DB-200		AT-210		VF-200 MS		007-210		G6

(Continued)

135

TABLE 3.13 (*continued*)

Restek	Phase Composition	Agilent/ J&W	Supelco	Alltech	SGE	Varian/ Chrompack	Phenomenex	Quadrex	Perkin-Elmer	USP Nomen-clature[a]
Rtx-50	50% Methyl- and 50% phenylpolysilox-ane	HP-17, HP-50+, DB-17, DB-17ht, DB-608	SP-2250, SPB-50	AT-50, AT-50 MS	BPX-50	VF-17 MS, CP Sil 24 CB	ZB-50	007-17	PE-17	G3
Rtx-65, Rtx-65 TG	35% Dimethyl- and 65% diphenylpoly-siloxane					TAP-CB		400-65HT, 007-65HT		G17
Rtx-225	50% Cyanopropyl-methyl- and 50% phenylmethyl-polysiloxane	HP-225, DB-225		AT-225	BP-225	CP Sil 43 CB		007-225	PE-225	G7, G19
Stabilwax, Rtx-WAX	Polyethylene glycol (PEG)	HP-20M, Inno Wax, DB-Wax, Carbowax 20M, HP-Wax, DB-Waxetr	Supelcowax-10, Carbowax PEG 20M	AT-Wax, Carbowax, AT-WAXMS	BP-20	CP Wax 52 CB	ZB-WAX	007-CW	PE-CW	G14, G15, G16, G20, G39
Stabilwax-DB	PEG for amines and basic compounds	CAM	Carbowax-Amine			CP Wax 51				
Stabilwax-DA	PEG for acidic compounds	HP-FFAP, DB-FFAP, OV-351	Nukol, SP-1000	AT-1000, FFAP	BP-21	CP Wax 58 CB		007-FFAP	PE-FFAP	G25, G35

Restek	Description								
Rtx-2330	90% Biscyanopropyl- and 10% cyanopropyl-phenylpolysiloxane		SP-2330, SP-2331, SP-2380	AT-Silar	BPX-70	CP Sil 84		007-23	G8
Rt-2560	Biscyanopropyl-polysiloxane		SP-2560						
Rt-bDEXm	Permethylated β-cyclodextrin	Cyclodex-b	b-DEX	Chiraldex-b	Cydex-b	CP-Cyclodextrin b			
Rt-TCEP	1,2,3-Tris(cyanoethoxy)propane		TCEP			CP-TCEP			
Rtx-440	Restek Exclusive!								
Stx-500	Phenylpoly-carborane–siloxane				HT-5				
Rtx-XLB	Proprietary phase	DB-XLB							
Column Features									
Integra-Guard	Built-in guard column	DuraGuard				EZ-Guard	Guardian		
MXT	Silcosteel-treated stainless steel	ProSteel	Metallon		Aluma-Clad	Ultimetal		Ultra-Alloy	

Source: Data from the Restek Corporation.
See Table 2.8.

Table 3.14 Columns by Application Cross-Reference

Restek	Application	Agilent/J&W	Supelco	Alltech	SGE	Varian/Chrompack	Phenomenex	Quadrex	PerkinElmer	USP Nomenclature[a]
Rtx-VMS, Rtx-VGC	EPA volatile organics methods 502.2, 524.2, 601, 602, 624, 8010, 8020, 8240, 8260	Restek Exclusive						Restek Exclusive		
Rtx-VRX, Rtx-502.2, Rtx-624, Rtx-Volatiles	EPA volatile organics methods 502.2, 524.2, 601, 602, 624, 8010, 8020, 8240, 8260	HP-624, HP-VOC, DB-624, DB-502.2, DB-VRX	VOCOL, SPB-624, SPB-624	AT-624		CP Sil 13 CB	OV-624	007-624, 007-502	PE-502	
Rtx-CLPesticides, Rtx-CLPesticides2, Stx-CLPesticides, Stx-CLPesticides2	EPA Methods 8081, 608, and CLP Pesticides	Restek Exclusive						Restek Exclusive		
Rtx-OPPesticides, Rtx-OPPesticides2	EPA Method 8141A OP Pesticides	Restek Exclusive						Restek Exclusive		
Rtx-5, Rtx-35, Rtx-50, Rtx-1701	EPA 608 Pesticides	HP-5, PAS-5, DB-5, DB-35, DB-608, HP-608, PAS-1701, DB-1701, DB-17, HP-50, HP-35	SPB-5, SPB-608, SPB-1701	AT-5, AT-35, AT-50, AT-1701, AT-Pesticides	BP-5, BP-10, BP-608	CP Sil 8 CB, CP Sil 19 CB	ZB-5, ZB-35, ZB-1701, ZB-50	007-2, 007-608, 007-17, 007-1701	PE-2, PE-608, PE-1701	G3

Rtx-2887	ASTM Test Method D2887	DB-2887	Petrocol 2887, Petrocol EX2887		CP-SimDist-CB	007-1-10V-1.0F
Rtx-1PONA	PONA analysis	HP-PONA, DB-Petro	Petrocol DH AT-Petro	BP1-PONA	CP Sil PONA CB	007-1-10V-0.5F
MXT-500 Sim Dist	Simulated distillation			HT-5		
Rtx-5 Amine	Amines and basic compounds		PTA-5			G50
FAMEWAX	Fatty acid methyl esters (FAMEs)		Omegawax			
Rtx-BAC, BAC2	Blood alcohol analysis	DB-ALC1, DB-ALC2				
Rtx-G27	Residual solvents in pharmaceuticals					G27
Rtx-G43	Residual solvents in pharmaceuticals		OVI-G43			G43
Rt-CW20M F&F	Fragrances and flavors	HP-20M, Carbowax 20M		BP-20M		007-CW
Rtx-TNT, Rtx-TNT2	Explosives	Restek Exclusive				Restek Exclusive
Rtx-Dioxin, Rtx-Dioxin2	Dioxins and furans	Restek Exclusive				Restek Exclusive

Source: Data from the Restek Corporation.

[a] See Table 2.8.

polysiloxane, the inference should not be drawn that the phases Rtx-50 and SP-2250 are identical, only that they are chemically similar and behave similarly under chromatographic conditions. Since each manufacturer has optimized column preparation for a specific stationary phase, column dimensions, and in some cases, an intended application of the column, slight differences in chromatographic behavior are to be expected. A manufacturer considers the steps involved in column preparation to be proprietary information.

Two types of stationary phases are most popular: the polysiloxanes and polyethylene glycol phases, as stated previously. Both types of phases may be characterized as having the necessary high viscosity and the capability for cross-linking and/or chemical bonding with fused silica. One should note the presence of more recent additions to the capillary column family: namely, specialty columns designed for selected EPA methods, chiral separations, and gas–solid chromatographic separations. Columns containing these phases are discussed in more detail in Section 3.6.

Polysiloxane Phases

Polysiloxanes are the most widely used stationary phases for packed and capillary column GC. They offer high solute diffusivities coupled with excellent chemical and thermal stability. The thorough review of polysiloxane phases by Haken (67) and the overview of stationary phases for capillary GC by Blomberg (68,69) are strongly recommended readings.

One measure of the polarity of a stationary phase is the cumulative value of its McReynolds constants, as described in Chapter 2. Because a variety of functional groups can be incorporated into the structure, polysiloxanes exhibit a wide range of polarities. Since many polysiloxanes are viscous gums and, as such, coat well on fused silica and can be cross-linked, they are ideally suited for capillary GC. The basic structure of 100% dimethylpolysiloxane can be illustrated as

$$\left[\begin{array}{cc} CH_3 & CH_3 \\ | & | \\ -Si-O-Si-O \\ | & | \\ CH_3 & CH_3 \end{array} \right]$$

Replacement of the methyl groups with another functionality enables polarity to be imparted to the polymer. The structure of substituted polysiloxanes in Figure 2.6 and those listed in Table 3.13 can be depicted by the following general representation:

$$\left[\begin{array}{c} R_1 \\ | \\ -Si-O- \\ | \\ R_2 \end{array} \right]_X \left[\begin{array}{c} R_3 \\ | \\ -Si-O- \\ | \\ R_4 \end{array} \right]_Y$$

where the R groups can be CH_3, phenyl, $CH_2CH_2CF_3$, or $CH_2CH_2CH_2CN$, and X and Y indicate the percentage of an aggregate in the overall polymeric stationary-phase composition, as described in Figure 3.6. In the case of the phase, DB-1301 or one of its chemically equivalents (6% cyanopropylphenyl–94% dimethylpolysiloxane, $R_1 = CH_2CH_2CH_2CN$, $R_2 =$ phenyl, and R_3 and R_4 are methyl groups; X and Y have values of 6 and 94%, respectively. For phases equivalent to 50% phenyl–50% methylpolysiloxane, R_1 and R_2 are methyl groups, while R_3 and R_4 are aromatic rings; X and Y are each equal to 50%. Additional polysiloxanes of various polarities have been described in the literature having polar functionalities of 4-(methylsulfonylphenyl) (70) and polyhydroxysubstitution (71) as well as medium-polarity phases having methoxy- (72) and silanol-terminated silarylene–siloxane goups (73).

Phase selectivity, which also affects resolution in a chromatographic separation, is governed by solute–stationary phase interactions, such as dispersion, dipole, acid–base, and hydrogen bond donors and acceptors. A column containing a polar stationary phase can display greater retention for a solute having a given polar functional group compared to other solutes of different functionality, while on a less polar stationary phase, the elution order may be reversed or altered to varying degrees under the same chromatographic conditions. Poole has investigated the selectivity equivalency of poly(dimethyldiphenylsiloxane) stationary phases for capillary GC (74). Several illustrations of how elution order is affected by stationary-phase selectivity are presented in Figures 3.24 and 3.25.

Polyethylene Glycol Phases

The most widely used non-silicon-containing stationary phases are the polyethylene glycols. They are commercially available in a wide range of molecular weights under several designations, such as Carbowax 20M and Superox-4. The general structure of a polyethylene glycol may be described as

$$HO-CH_2-CH_2-(-O-CH_2-CH_2-)_n-O-CH_2-CH_2-OH$$

The popularity of polyethylene glycols stems from their unique selectivity and high polarity as a liquid phase. Unfortunately, they do have some limitations. A characteristic of Carbowax 20M, for example, is its rather low upper temperature limit of approximately $225°C$ and minimum operating temperature of $60°C$. In addition, trace levels of oxygen and water have adverse effects on most liquid phases, but particularly so with Carbowax 20M, where they accelerate the degradation process of the phase. Verzele and co-workers have attempted to counteract these drawbacks by preparing a very high molecular weight polyethylene glycol, Superox-4 (75). Other successful attempts include free-radical cross-linking and bonding, which are discussed in the next section.

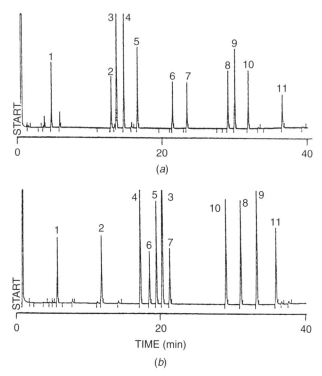

Figure 3.24 Chromatograms of the separation of nitro-containing compounds on 10 m × 0.53 mm i.d. column with (a) CP Sil 19 CB and (b) CP Sil 8 CB as stationary phase; column conditions: 50°C (3 min) at 6°C/min to 250°C, detector FID, 24 cm/s He. Solutes: (1) nitropropane, (2) 2-nitro-2-methyl-1-propanol, (3) n-dodecane, (4) nitrobenzene, (5) o-nitrotoluene, (6) 2-nitro-2-methyl-1,3-propanediol, (7) 2-nitro-2-ethyl-1,3-propanediol, (8) p-nitroaniline, (9) p-nitrobenzyl alcohol, (10) o-nitrodiphenyl, and (11) 4-nitrophthalimide. The stationary phases, CP Sil 19 CB and Cp Sil 8 CB are chemically similar to OV-1701 and OV-5, respectively. (From ref. 146.)

Cross-Linked Versus Chemically Bonded Phase

The practice of capillary GC has been enriched by the advances made in the immobilization of a thin film of a viscous stationary phase coated uniformly on the inner wall of fused-silica tubing. At present, two pathways are employed for the immobilization of a stationary phase: free-radical cross-linking and chemical bonding. By immobilizing a stationary phase by either approach, the film is stabilized and is not disrupted at elevated column temperatures or during temperature programming. Thus, less column bleed and higher operating temperatures can be expected with a phase of this nature, a consideration especially important in GC-MS. A column containing an immobilized stationary phase is also recommended for on-column injection and large volume injectors/cool-on-column inlets where large aliquots of solvent are injected without dissolution of the stationary phase. Similarly, a column

1   Methanol
2.   Methyl formate
3.   Ethanol
4.   Acetone
5.   2-propanol
6   Ethyl formate
7   1,1-Dichlorocthylenc
8.   Methylene chloride
9   Methyl acelate
10.   1-Propanol
11.   Trans 1,2-Dichloroethylene
12.   1,1-Dichloroethane
13.   2-Butanone
14.   Sec-Butanol
15.   Hexane

16.   Ethyl acetate
17.   Chloroform
18.   Tetrahydrofuran
19.   Isobutanol
20.   2-Methoxyethanol
21.   1,2-Dichloroethane
22.   1,1,1-Trichloroethane
23.   Isopropyl acetate
24.   n-Butanol
25.   Benzene
26.   Carbon Tetrachloride
27.   2-Nitropropane
28.   Trichloroethylene
29.   1,4-Dioxane
30.   2-Ethoxyethanol
    (cellosolve)

31.   n-Propyl acetate
32.   4-Methyl-2-pentanone
33.   Isoamyl alcohol
34.   Dimethylformamide
35.   Toluene
36.   Isobutyl acetate
37.   2-Hexanone
38.   Mesityl oxide
39.   Tetrachloroethene
40.   n-butyl acetate
41.   Diacetone alcohol
42.   Chlorobenzene
43.   S-Methyl-2-hexanone
44.   Ethyl benzene
45.   m-Xylene

46.   p-Xylene
47.   Isoamyl acetate
48.   Cyclohexanol
49.   Styrene
50.   o-Xylene
51.   1,1,2,2-Tetrachloroethane
52.   2-Ethoxyethyl acetate
53.   Butyl cellosolve
54.   n-Amyl acetate
55.   2-Methylcyclohexanol
56.   1,2-Dichlorobenzene
57.   2-Methylphenol
58.   3-Methylphenol
59.   4-Methylphenol

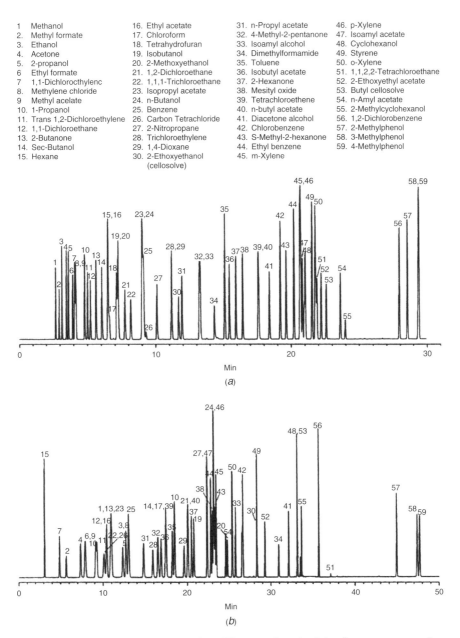

Figure 3.25   Chromatograms illustrating differences in selectivity for components of an industrial solvent mixture with (*a*) Equity-1 stationary phase; (*b*) Supelcowax stationary phase. In each case, column dimensions 30 m × 0.32 mm i.d., 1.0 μm; temperature conditions 35°C (8 min) at 4°C/min to 130°C, detector FID, 25 cm/s He, split injection (200:1). (Reprinted with permission of Sigma-Aldrich Co.)

having an immobilized phase can be backflushed to rinse contamination from the column without disturbing the stationary phase (Section 3.8).

Cross-Linking of a Stationary Phase.   The ability of a polymer to cross-link is highly dependent on its structure. The overall effect of cross-linking is that the molecular weight of the polymer steadily increases with the degree of cross-linking, leading to branched chains, until eventually a three-dimensional rigid network is formed. Since the resulting polymer is rigid, little opportunity exists for the polymer chains to slide past one another, thereby increasing the viscosity of the polymer. Upon treatment of a cross-linked polymer with solvent, the polymer does not dissolve; rather, a swollen gel remains behind after decantation of the solvent. Under the same conditions, an uncross-linked polymer of the same structure would dissolve completely. In summary, the dimensional stability, viscosity, and solvent resistance of a polymer are increased as a result of cross-linking. The mechanism for cross-linking 100% dimethylpolysiloxane is described below, where R· and (gamma radiation) are free-radical initiators:

$$
\begin{array}{ccccc}
& CH_3 & & CH_3 & CH_3 \\
& | & & | & | \\
-\!Si\!-\!O\!- & & -\!Si\!-\!O\!- & & -\!Si\!-\!O\!- \\
& | & & | & | \\
& CH_3 & & \bullet CH_2 & CH_2 \qquad +2RH \\
& & R\bullet & & | \\
& & +\ or\ \rightarrow & & CH_2 \qquad or \\
& & \gamma & & | \\
& & & & \rightarrow\!-\!Si\!-\!O\!-\quad H_2 \\
& CH_3 & & CH_3 & | \\
& | & & | & CH_3 \\
-\!Si\!-\!O\!- & & -\!Si\!-\!O\!- & & \\
& | & & | & \\
& CH_3 & & \bullet CH_2 & \\
\end{array}
$$

The sequence of photographs presented in Figure 3.26 permits a helpful visualization of the cross-linking process for several stationary phases, 100% dimethylpolysiloxane (SE-30) and trifluoropropylmethylpolysiloxane, as a function of increasing degree of cross-linking by ^{60}Co gamma radiation (76). The conversion of DC200 (a silicone oil) and OV-101 (a polysiloxane fluid) to cross-linked gel versions as a function of radiation dosage is illustrated in Figure 3.26c and d.

Madani et al. provided the first detailed description of capillary columns where polysiloxanes were immobilized by hydrolysis of dimethyl- and diphenylchlorosilanes (77,78). Interest increased when Grob found that the formation of cross-linked polysiloxanes resulted in enhanced film stability (79). Blomberg et al. illustrated in situsynthesis of polysiloxanes with silicon tetrachloride as a precursor, followed by polysiloxane solution (80,81). Since then, various approaches to cross-linking have been investigated. These include chemical additives such as organic peroxides (82–90), azo compounds (65,91,92), ozone (93), and gamma radiation (43,94–97). Several different peroxides have been evaluated, dicumyl peroxide being the most popular. However, peroxides can generate polar decomposition products that remain in the immobilized film of stationary phase. Moreover, oxidation may also occur, which increases the polarity and decreases the thermal stability of a column. These adverse effects are eliminated with azo species as free-radical initiators. Lee et al. have cross-linked a wide range of stationary

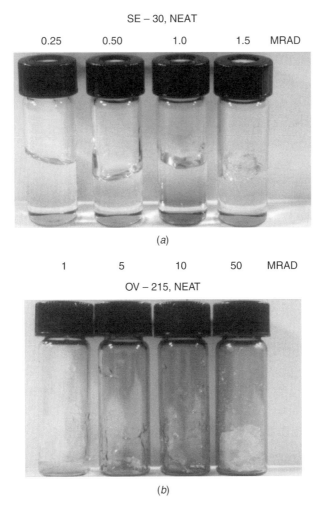

Figure 3.26 Effect of gamma radiation on degree of cross-linking of (*a*) SE-30 (poly-dimethylsiloxane; (*b*) OV-215 (trifluoropropylmethylpolysiloxane; (*c*) DC 200 (a silicone oil) and (*d*) conversion of OV-101 (a polydimethylpolysiloxane fluid) to a gum similar to OV-1. [(*a*, *b*) From ref. 76, reproduced from the *Journal of Chromatographic Science*, by permission of Preston Publications, a division of Preston Industries, Inc.] (*Continued*)

Figure 3.26   (*continued*)

phases, from nonpolar to polar, in their studies using azo-*tert*-butane and other azo species (48,91,92). If an azo compound or a peroxide exists as a solid at room temperature, the agent is spiked directly into the solution of the stationary phase used for coating the column. On the other hand, for free-radical initiators that are liquid at ambient temperature, the column is first coated with stationary phase, then saturated with vapors of the reagent (47,82).

Gamma radiation from a ^{60}Co source has also been used by Schomburg et al. (43), Bertsch et al. (94), and Hubball and co-workers (95–97) as an effective technique for cross-linking polysiloxanes. In a comparative study of gamma

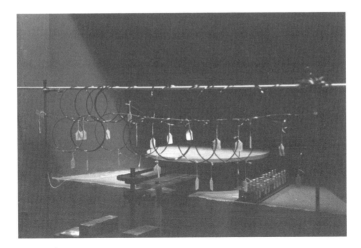

Figure 3.27   Capillary columns positioned for irradiation by ^{60}Co in the University of Massachusetts Lowell Radiation Facility; packings positioned for irradiation and cross-linking of the stationary phase are evident in the lower right-hand corner.

radiation with peroxides, Schomburg et al. (43) noted that each approach immobilized polysiloxanes, but that the formation of polar decomposition products is avoided with radiation. Radiation offers the additional advantages that the cross-linking reaction occurs at room temperature and columns can be tested both before and after the immobilization of the stationary phase. The positioning of a group of capillary columns and packings in the gamma irradiation facility at the University of Massachusetts Lowell is displayed in Figure 3.27.

Not all polysiloxanes can be cross-linked directly or readily. The presence of methyl groups facilitates cross-linking. Consequently, the nonpolar siloxanes exhibit very high efficiencies and high thermal stability. However, as the population of methyl groups on a polysiloxane phase decreases and the groups are replaced by phenyl or more polar functionalities, cross-linking a polymer becomes more difficult. The incorporation of vinyl or tolyl groups into the synthesis of a polymer tailored for use as a stationary phase for capillary GC overcomes this problem. Lee et al. (88,91) and Blomberg et al. (89,90) have successfully synthesized and cross-linked stationary phases of high phenyl and high cyanopropyl content that also contain varying amounts of these free-radical initiators. Colloidal particles have also been utilized to stabilize cyanoalkyl stationary-phase films for capillary GC (98). Recently, favorable thermal stability and column inertness were obtained by a binary cross-linking reagent, a mixture of dicumyl peroxide and tetra(methylvinyl)cyclotetrasiloxane (99).

Developments in the cross-linking of polyethylene glycols have been slower in forthcoming, although successes have been reported. Immobilization of this phase by the following procedures increases its thermal stability and its compatibility and tolerance for aqueous solutions. De Nijs and de Zeeuw (100) and Buijten et al. (101)

immobilized a polyethylene glycol in situ, the latter group with dicumyl peroxide and methyl(vinyl)cyclopentasiloxane as additives. Etler and Vigh (102) used a combination of gamma radiation with organic peroxides to achieve immobilization of this polymer, whereas Bystricky selected a 40% solution of dicumyl peroxide (103). George (104) and Hubball (105) have cross-linked polyethylene glycol successfully using radiation; in Figure 3.28, an array of vials of Carbowax 20M shown after receiving various dosages of gamma radiation indicates that cross-linking has been achieved (vial D). The chromatographic separation of a cologne in Figure 3.29 was generated on a capillary column containing Carbowax 20M cross-linked by gamma radiation and indicates acceptable thermal stability to 280°C. Horka and colleagues described a procedure for cross-linking Carbowax 20M with pluriisocyanate reagents (106). Recently, thermally bondable polyethylene glycols and polyethyleneimines have

CARBOWAX 20M

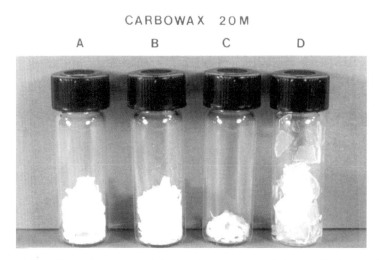

Figure 3.28 Effect of gamma radiation on degree of cross-linking of Carbowax 20M.

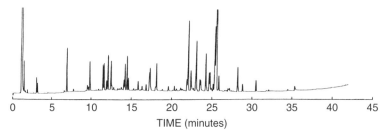

Figure 3.29 Chromatogram illustrating the separation of a cologne sample on a capillary column (15 m × 0.32 mm i.d., 0.25-μm film) containing Carbowax 20M cross-linked by gamma radiation; column conditions: 40°C (2 min) at 6°C/min to 280°C, detector FID, 25 cm/s He.

been popular phases and yield chromatographic selectivity similar to that of the traditional polyethylene glycols. Despite these efforts, the upper temperature limit of polyethylene glycol columns generally remains below 300°C.

Chemical Bonding

Since the early 1990s, column manufacturers have devoted extensive resources to acquiring technology for the development of chemically bonding a stationary-phase film to the inner wall of a fused-silica capillary. As the term suggests, an actual chemical bond is formed between the fused silica and the stationary phase. The foundation of this procedure was first reported by Lipsky and McMurray (107) in their investigation of hydroxy-terminated polymethylsilicones, and was later refined by the work of Blum et al. (108–112), who employed OH-terminated phases for the preparation of inert, high-temperature stationary phases of varying polarities. The performance of hydroxy-terminated phases has also been evaluated by Schmid and Mueller (113) and Welsch and Teichmann (114). Other published studies include the behavior of hydroxyl phases of high cyanopropyl content by David et al. (115) and of trifluoropropylmethylpolysiloxane phases by Aichholz (116).

In the chemical bonding approach to stationary-phase immobilization, a capillary column is coated in the conventional fashion with an OH-terminated polysiloxane and then temperature programmed to an elevated temperature, during which time a condensation reaction occurs between the surface silanols residing on the fused-silica surface and those of the phase. It is important to note here that both deactivation and coating are accomplished in a single-step process and result in the formation of a Si–O–Si bond more thermally stable than the Si–C–C–Si bond created via cross-linking. Cross-linking of the stationary phase is not a necessary requirement, but if a stationary phase contains a vinyl group (or another free-radical initiator), cross-linking can occur simultaneously. Phases that cannot be cross-linked during the bonding process can be cross-linked afterward with an azo compound, for example. After observing the increased inertness and thermostability of OH-terminated phases, Grob commented that they might reflect a "revolution in column technology" (117).

Since the mid-1980s, immobilization of polysiloxanes and polyethylene glycols is no longer a subject of rapid advancements reported in the literature; a procedural blend of polymeric synthesis, cross-linking, and/or chemically bonding is utilized by column manufacturers today, as the fixation of these stationary phases via cross-linking and/or chemical bonding for capillary GC is now a well-defined and mature technology. A capillary column containing such a stationary phase is a result of elegant pioneering efforts by people such as M. L. Lee, L. Blomberg, K. Grob and his family members, G. Schomburg and their colleagues, and many, many others too numerous to mention here. Attention has now shifted to such areas as polysiloxanes of high cyano content, silphenylene(arylene) phases for GC-MS, MS-grade phases, stationary-phase selectivity tuned or optimized for specific applications, multidimensional chromatographic techniques, and immobilization of chiral stationary phases, which are discussed in Section 3.6.

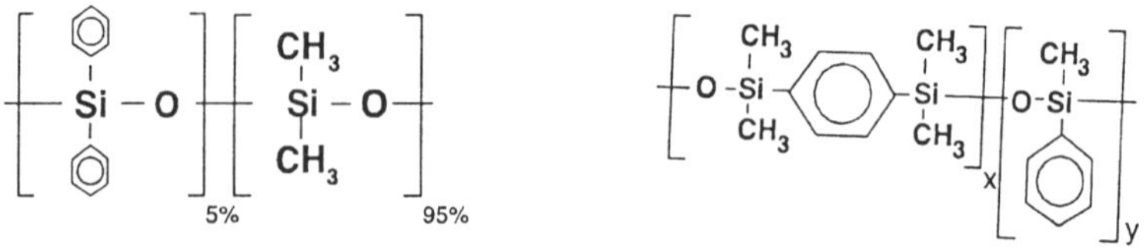

5% Phenylmethylpolysiloxane   **5% Phenylpolysilphenylene-siloxane**

Figure 3.30   Structures of 5% phenyl–95% methylpolysiloxane and 5% phenylpoly-silphenylene–siloxane.

MS-Grade Phases Versus Polysilarylene or Polysilphenylene Phases

Many column manufacturers offer what has become known as *MS columns*: columns that generate lower bleed than do regular or conventional polysiloxane equivalents. Lower bleed is highly desirable in GC-MS analyses because complications such as bleed ions, misidentifications of compounds, and errors in quantitation, to name few, are avoided. A MS-grade column may contain (1) a higher-molecular-weight polymer obtained by a fractional procedure of the corresponding starting stationary-phase version (2) a polymer resulting from cross-linking of a higher-molecular-weight fraction of the starting polymer, (3) a "cross-bonded" polymer resulting from the condensation reaction of a hydroxyl-terminated polymer of either a conventional polysiloxane or a higher molecular fraction with fused silica where cross-linking may or may not have occurred (as discussed in the preceding section), or (4) the stationary phase may be a polysilarylene–siloxane, often referred to as an *arylene phase*, or as a polysilphenylsiloxane, often referred to as a *phenylene phase*. The latter phases are inherently more thermally stable because of the presence of aromatic rings in the polymer chain, as depicted in Figure 3.30, and represent an upgrade in thermal performance over the corresponding polysiloxane counterpart, but one may also notice slight differences in selectivity due to the different chemistry employed both in the deactivation procedure and also in a possible synthesis of the phase itself. Thus, cross-reference column charts should be examined carefully when comparing columns offered by various vendors when selecting a capillary column for an application requiring a low column bleed level.

Sol-Gel Stationary Phases

Sol-gel is basically a synthetic glass with ceramiclike properties. The processing consists of hydrolysis and condensation of a metal alkoxide (i.e., tetraethoxysilane) to form a glassy material at room temperature. Further modification of this material with a polymer (stationary phase) is used to prepare phases for capillary columns; there has been keen interest in this process (118–123). The final sol-gel product retains the properties of the polymer as well as the properties of the sol-gel component. The sol-gel material is able to covalently bond to fused silica,

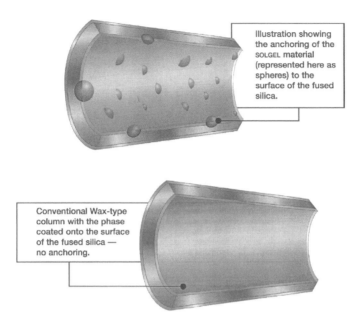

Figure 3.31 Comparison of cross section of a SolGel-WAX column with a conventional WAX-type column. (Courtesy of SGE Analytical Sciences.)

yielding a strong bond, which results in better thermal stability and less column bleed. In addition, the molecular weight of the stationary phase is stabilized via end-capping chemistry, providing protection from degradation and potential further condensation. At the present time, two gel-sol phases have been developed: SolGel-1 ms, derived from 100% dimethylpolysiloxane, and SolGel-WAX, which has polyethylene glycol in the matrix. A cross-sectional view of a SolGel-WAX column along with a corresponding view of a conventionally coated capillary is presented in Figure 3.31; an application of a separation with this type of column appears in Figure 3.32.

Phenylpolycarborane–Siloxane Phases

This classification of phases can be traced back to the previous use of the carborane-type phase Dexsil, which was widely utilized as a stationary phase in packed columns for high-temperature separations because of its excellent thermal stability. The carborane network has been incorporated into the backbone of phenylpolysiloxane phases having either a 5 or 8% phenyl content, providing unique selectivity for selected applications. The structure of this modified polysiloxane is presented in Figure 3.33. A capillary column containing this stationary phase exhibits high selectivity for difficult-to-separate Aroclor 1242 congeners because of the carborane functionality in the polysiloxane polymer. An example of this is shown in Figure 3.34, where the carborane phase interacts preferentially with ortho-substituted PCB congeners (i.e., congeners 28 and 31).

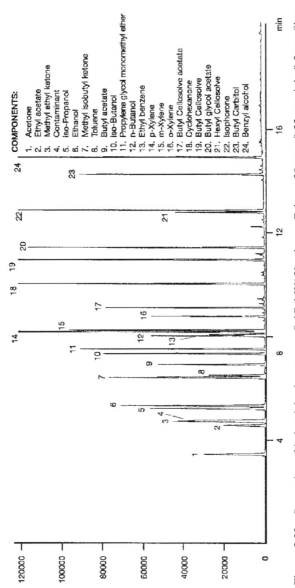

COMPONENTS:

1. Acetone
2. Ethyl acetate
3. Methyl ethyl ketone
4. Contaminant
5. iso-Propanol
6. Ethanol
7. Methyl isobutyl ketone
8. Toluene
9. Butyl acetate
10. iso-Butanol
11. Propylene glycol monomethyl ether
12. n-Butanol
13. Ethyl benzene
14. p-Xylene
15. m-Xylene
16. o-Xylene
17. Butyl Cellosolve acetate
18. Cyclohexanone
19. Butyl Cellosolve
20. Butyl glycol acetate
21. Hexyl Cellosolve
22. Isophorone
23. Butyl Carbitol
24. Benzyl alcohol

Figure 3.32 Separation of industrial solvents on a SolGel-WAX column. Column 30-m × 0.32 mm i.d., 0.5-μm film; temperature conditions 35°C (3 min) at 15°C/min to 230°C, detector FID, 1.84 mL/min, 30 cm/s He, split injection (83:1) 240°C. (Courtesy of SGE Analytical Sciences.)

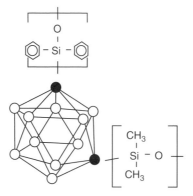

Figure 3.33   Structure of a phenylpolycarborane–siloxane stationary phase.

3.6   SPECIALTY COLUMNS

EPA Methods

Column manufacturers have responded to the increasing analytical and environmental demands for capillary columns for use in EPA 500 series, EPA 600 series, and EPA 8000 series methods. To simplify column selection for a given EPA method by a user, a column is designated by the alphabetical or numerical tradename of the manufacturer followed by the EPA method number, as shown in Table 3.13. These columns have been configured in length, inner diameter, film thickness, and stationary-phase composition for optimized separation of the targeted compounds under the chromatographic conditions specified in a particular method. Another factor relating to column dimensions considered by manufacturers is the compatibility of thicker film columns with methods that stipulate purge-and-trap sampling by eliminating the need for cryogenic solute focusing prior to chromatographic separation. EPA methods and corresponding recommended column configurations are discussed at length in Section 3.7.

Chiral Stationary Phases

Capillary columns that have a chiral stationary phase are used for the separation of optically active isomers or enantiomers: species that have the same physical and chemical properties with the exception of the direction in which they rotate plane-polarized light. Enantiomers may also have different biological activity, and therefore enantiomeric separations are important in the food, flavor, and pharmaceutical areas. A chiral stationary phase can recognize differences in the optical activity of solutes to a varying extent, whereas common stationary phases do not. Chromatograms illustrating the separation of enantiomers present in lemon and rosemary oils are given in Figure 3.35.

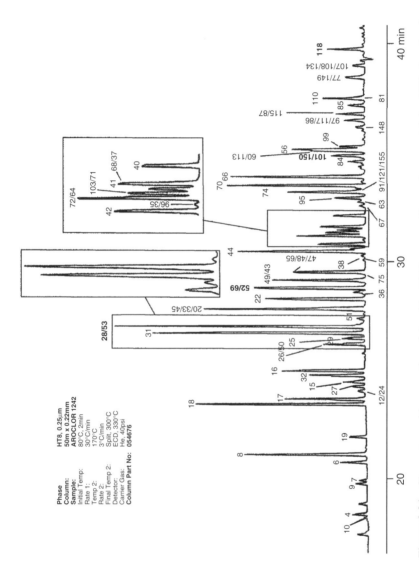

Figure 3.34 Chromatographic separation of Arochlor 1242 on a capillary column containing HT8 phenyl poly-carborane–siloxane as the stationary phase; column 50 m × 0.22 mm i.d., 0.25-μm film; temperature conditions 80°C (2 min) at 30°C/min to 170°C, then 3°C/min to 300°C, detector ECD, 1.84 mL/min, 40 psi He, split injection. (Courtesy of SGE Analytical Sciences.)

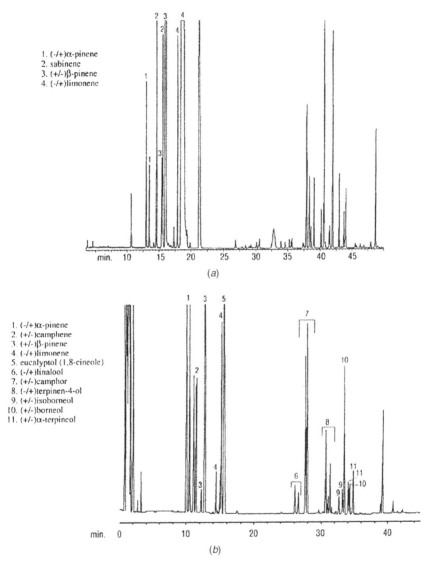

1. (-/+)α-pinene
2. sabinene
3. (+/-)β-pinene
4. (-/+)limonene

(a)

1. (-/+)α-pinene
2. (+/-)camphene
3. (+/-)β-pinene
4. (-/+)limonene
5. eucalyptol (1,8-cineole)
6. (-/+)linalool
7. (+/-)camphor
8. (-/+)terpinen-4-ol
9. (+/-)isoborneol
10. (+/-)borneol
11. (+/-)α-terpineol

(b)

Figure 3.35 Chiral analysis of (a) lemon oil and (b) rosemary oil. Column 30 m × 0.32 mm i.d., 0.25-μm film, Rt-βDEXsm; temperature conditions: 40°C (1 min) at 2°C/min to 200°C (hold 3 min), detector FID, 80 cm/s He, split injection. (Courtesy of the Restek Corporation.)

Cyclodextrins (Alpha, Beta, and Gamma). Most of the earlier chiral phases have limited thermal stability. By chemically bonding a chiral stationary phase to a polysiloxane, the upper temperature limit can be extended. Chirasil-Val is perhaps the most famous stationary phase in this category. However, recent work has employed cyclodextrins as the key chiral recognition component in stationary phases. The mechanism of separation (or enantiomeric selectivity) is based on the formation of solute–cyclodextrin complexes occurring in the barreled-shaped opening of cyclodextrin and can be manipulated by varying the size of the opening of the cyclodextrin ring as well as by the alkyl substitution pattern on the ring. Structures of the most widely used cyclodextrins, which differ in the number of glucose units (6, 7, and 8) in the structure, are shown in Figure 3.36. They therefore differ in the diameter of the toroidal geometry and available surface within the cavity. Although cyclodextrins can be used as a stationary phase (124,125), the current practice is either to place them in solution with a polysiloxane (126) or to immobilize them by bonding to a polysiloxane such as a cyanopropyldimethylpolysiloxane (127). The chromatographic principles underlying chiral separations by GC have been reviewed by Hinshaw (128).

Most column vendors offer a diverse product line of chiral columns with application notes available online or in hard-copy format. Perhaps the most extensive array of columns are the family of Chiraldex columns manufactured by Advanced Separation Technologies, Inc., which has published a helpful handbook which provides an understanding of structural relationships, reversals in elution order and underlying mechanistic theories, as well as providing an extensive number of chiral separations to assist in making the correct column choice (129). A list of chemical classes of compounds and functional groups amendable to chiral separations is provided in Table 3.15. Ding and co-workers reported that room-temperature ionic liquids can also exhibit chiral properties and demonstrated enantiomeric separations (130); studies of this nature can greatly extend the realm of chiral separations.

Gas–Solid Adsorption Capillary Columns: PLOT Columns

Gas–solid adsorption capillary columns are also referred to as *PLOT* (*porous-layer open-tubular*) *columns*. A PLOT column consists of fused-silica capillary tubing in which a layer of adsorbent lines the inner wall in place of a liquid phase. An early use of a PLOT column was reported by de Nijs (131), who prepared a fused-silica column coated with submicron particles of aluminum oxide for the analysis of light hydrocarbons. De Zeeuw et al. (132) subsequently prepared many types of PLOT columns, including a column with a 10- to 30- m layer of a porous polymer of the Porapak Q type (styrene–divinylbenzene) by in situ polymerization of a coating solution. Because of improved column technology, adsorbent-type stationary-phase particles can now be bonded very effectively to the inner surface texture of fused silica, eliminating the need for particulate traps and the possibility of column particles accumulating in a FID jet or in a TCD cell, although column vendors do offer short lengths of guard columns installed after a PLOT column. A highly informative review of PLOT columns, including their preparation and applications, by Ji et al. is strongly recommended reading (133). PLOT columns are now commercially available which have the counterparts of adsorbents used in GSC as the stationary phase now bonded to an inner wall of fused silica. Many laboratories are transferring more and more of their analytical methods once performed with packed gas–solid chromatographic columns over to the corresponding PLOT counterpart.

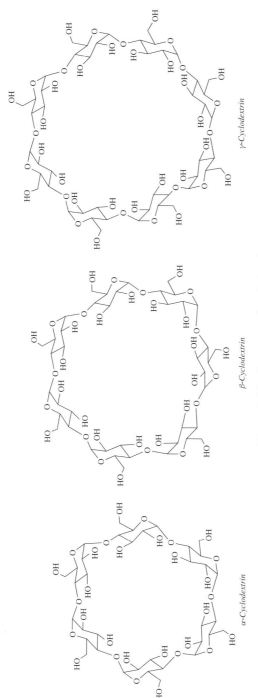

Figure 3.36 Structures of cyclodextrins.

TABLE 3.15   General Classes of Compounds Separated on Selected Chiraldex Capillary Columns

The following abbreviations are utilized below:

TA Trifluoroacetyl	DP Dipropionyl
(2,6-di-*O*-pentyl-3-trifluoroacetyl)	(2,3-di-*O*-propionyl-6-*t*-butyl silyl)
DM Dimethyl	DA Dialkyl
(2,3-di-*O*-methyl-6-*t*-butyl silyl)	(2,6-di-*O*-pentyl-3-methoxy)
PN Propionyl	BP Butyl
(2,6-*O*-pentyl-3-propionyl)	(2,6-*O*-pentyl-3-butyl)
PH *S*-Hydroxylpropyl	PM Permethyl
[(*S*)-2-hydroxypropyl methyl ether]	(2,3,6-tri-*O*-methyl)

The prefixes A, B, and G refer to α, β, and γ cyclodextrins, respectively.

A-PH	Smaller linear and saturated amines, alcohols, carboxylic acids, and epoxides
B-PH	Most structural types of compounds, including linear and cyclic amines and alcohols, carboxylic acids, lactones, amino alcohols, sugars, bicyclics, epoxides, haloalkanes, and more
G-PH	Cyclic and bicyclic diols and other larger compounds, including steroids and carbohydrates
A-DA	Smaller cyclic amines, alcohols, and epoxides (inquire regarding availability)
B-DA	Nitrogen heterocyclics, heterocyclics, some bicyclics, and epoxides, lactones, aromatic amines, sugars, and amino acid derivatives
G-DA	Aromatic amines containing two or more rings, large cyclic diols, some heterocyclics, multiring compounds, or compounds with bulky substituents
A-TA	Smaller alcohols, amino alcohols, amino alkanes, and diols
B-TA	Broad range of alkyl alcohols, halo acid esters, amino alkanes, halocycloalkanes, certain lactones, diols, alkyl halides, and furan and pyran derivatives
G-TA	>350 pairs chiral alcohols, diols, polyols, hydrocarbons, lactones, amine alcohols, halocarboxylic acid esters, homologous series, furan and pyran derivatives, epoxides, glycidyl analogs, and haloepihydrins
G-PN	Epoxides, higher alcohols >C4, and lactones
G-BP	Amino acids, certain primary amines, and furans
G-DM	Selectivity similar to PM and PH, but with shorter retention times, and greater resolution; especially applicable to aliphatic, olefenic, and aromatic enantiomers
B-PM	Acids, alcohols, barbitals, diols, epoxides, esters, hydrocarbons, ketones, lactones, and terpenes
B-DM	Selectivity similar to that of PM and PH, but with shorter retention times and greater resolution; especially applicable to aliphatic, olefenic, and aromatic enantiomers
B-DP	Aromatic and aliphatic amines and some aliphatic and aromatic esters
G-DP	Aromatic and aliphatic amines and some aliphatic and aromatic esters

Several separations on PLOT columns are displayed in Figures 3.37 and 3.38. The selection of adsorbents currently available in a PLOT column format includes aluminum oxide–KCl, aluminum oxide–sodium sulfate, molecular sieves, alumina, graphitized carbon, and porous polymers. These columns are intended to be direct

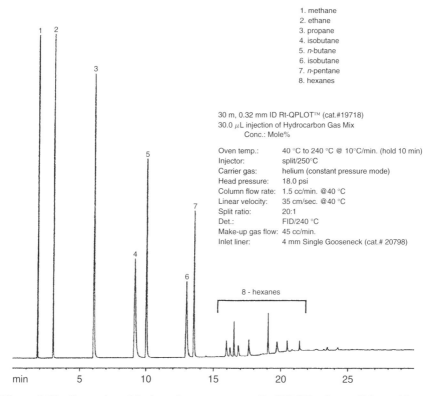

1. methane
2. ethane
3. propane
4. isobutane
5. n-butane
6. isobutane
7. n-pentane
8. hexanes

30 m, 0.32 mm ID Rt-QPLOT™ (cat.#19718)
30.0 μL injection of Hydrocarbon Gas Mix
Conc.: Mole%

Oven temp.:	40 °C to 240 °C @ 10°C/min. (hold 10 min)
Injector:	split/250°C
Carrier gas:	helium (constant pressure mode)
Head pressure:	18.0 psi
Column flow rate:	1.5 cc/min. @40 °C
Linear velocity:	35 cm/sec. @40 °C
Split ratio:	20:1
Det.:	FID/240 °C
Make-up gas flow:	45 cc/min.
Inlet liner:	4 mm Single Gooseneck (cat.# 20798)

Figure 3.37   Separation of hydrocarbon gases on an Rt-QPLOT column. Column 30 m × 0.32 mm i.d., Rt-QPLOT; temperature conditions 40°C at 10°C/min to 240°C (hold 10 min), detector FID, 18 psi He, 35 cm/s, split injection. (Courtesy of the Restek Corporation.)

replacements for a packed column containing the same adsorbent, and feature faster regeneration of the adsorbent. Applications of various PLOT columns are listed in Table 3.16. which clearly indicates diversity in many areas of chromatographic significance. De Zeeuw and Barnes published the results of a selectivity study with various PLOT columns with the goal of faster time of analysis (134); several interesting photomicrographs are shown in Figure 3.39. In Appendix A, separations achieved with gas–solid chromatographic columns are presented to assist anyone who wishes to cross-reference a specific adsorbent with the appropriate stationary phase in the PLOT column format in Table 3.16.

3.7   CAPILLARY COLUMN SELECTION

Practical Considerations of Column Diameter, Film Thickness, and Column Length

Guidelines for the selection of column diameter, film thickness of stationary phase, and length will now be established based on practical gas chromatographic considerations.

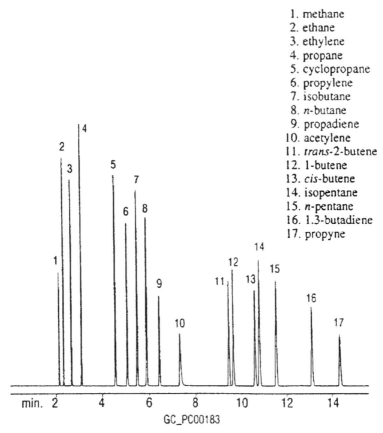

1. methane
2. ethane
3. ethylene
4. propane
5. cyclopropane
6. propylene
7. isobutane
8. *n*-butane
9. propadiene
10. acetylene
11. *trans*-2-butene
12. 1-butene
13. *cis*-butene
14. isopentane
15. *n*-pentane
16. 1.3-butadiene
17. propyne

GC_PC00183

Figure 3.38   Separation of refinery gases on Rt-Alumina PLOT column. Column 50 m ×
0.53 mm i.d., Rt_QPLOT; temperature conditions 5°C at 10°C/min to 120°C (hold 5 min),
detector FID, He 37 cm/s, split injection, FID. (Courtesy of the Restek Corporation.)

Column Diameter

1. Sample capacity increases as column diameter increases. Samples that have
   components present in the same concentration range can be analyzed on a
   column of any diameter. The choice depends on the resolution required. In
   general, the sample capacity of any capillary column is proportional to the
   square of the column radius.
2. For complex samples, select a column that has the smallest diameter and
   sample capacity compatible with the concentration range of the sample components.
3. Samples whose components differ widely in concentration should be analyzed
   on a column of larger i.d. (>0.25 mm) to avoid overload of the column by
   solutes of higher concentration.

TABLE 3.16 PLOT Columns and Applications

Column	Application
Alumina	General-purpose alumina column; hydrocarbon analysis, C1–C5 hydrocarbons; analysis of ethylene, propylene, and butanes.
Alumina-KCl	Alumina deactivated with KCl; least polar alumina; C1–C8 hydrocarbon isomers; lowest retention of olefins relative to comparable paraffin. Quantitation of dienes, especially propadiene and butadiene in presence of ethylene and propylene.
Alumina-S	Alumina deactivated with sodium sulfate; midrange polarity of alumina phases; general-purpose alumina; excellent for resolving acetylene from butane and propylene from isobutane.
Molecular Sieve 5A	Permanent and noble gases [i.e., argon–oxygen separation (may necessitate a thick film)]; gas purity analysis.
Molecular Sieve 13X	Separation of hydrogen, oxygen, nitrogen, methane, and CO.
Q-type PLOT	Dinylbenzene–polystyrene copolymer; gases and voltalile organics, polar solvents, alcohols, nonpolar hydrocarbons, C1–C3 isomers, alkanes to C12, carbon dioxide, methane, air–CO, water, sulfur compounds, solvents.
U-type PLOT	Divinylbenzene/ethylene glycol dimethylacrylate copolymer; polar volatiles, nitriles, nitro comounds, alcohols, aldehydes, ethane–ethylene, C1–C7 hydrocarbons, amines, solvents.
S-type PLOT	Divinylbenzene vinyl pyridine polymer; light gases in ethylene and propylene, ketones, and esters.
Carboxen 1006 PLOT	Carbon molecular sieve; permanent gases, C1, C2, and C3 light hydrocarbons; separation of formaldehyde–water–methanol mixtures.
Carboxen 1010 PLOT	Carbon molecular sieve; separation of hydrogen, nitrogen, CO, methane, carbon dioxide, and C2 and C3 hydrocarbons.
CarbonPLOT	Bonded, monolithic carbon layer; C1–C5 hydrocarbons, carbon dioxide, air–CO, trace acetylene in ethylene, methane

Source: 2005 catalogs of Agilent Technologies, Restek Corporation, and Supelco.

4. The selection of column i.d. may be based on the type of sample inlet system. Generally, a 0.25- or 0.32-mm-i.d. column may be used for split and splitless injections, 0.32 mm i.d. for splitless and on-column injections, and 0.53 mm i.d. for direct injections.

5. Capillary columns of 0.18 and 0.25 mm i.d. should be used for GC-MS systems, because the lower flow rates with these columns will not exceed the limitations of the vacuum system.

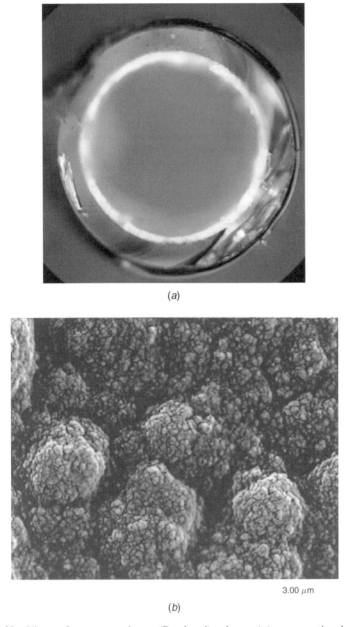

(a)

3.00 μm

(b)

Figure 3.39 Views of a porous polymer (Porabond) column. (a) cross-sectional view; (b) helical view and (c) sideview of Porabond column prepared in situ; (d) separation of a series of solvents. (Courtesy of Jaap de Zeeuw, Varian BV.)

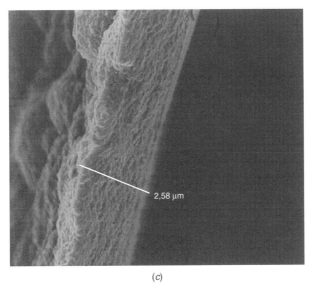

(c)

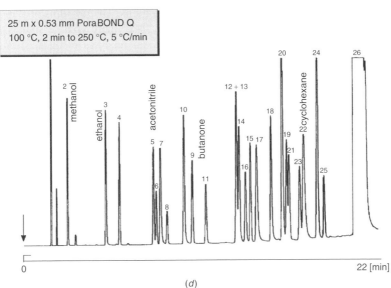

(d)

Figure 3.39   (*continued*)

6. Fast capillary columns (0.10 mm i.d.) are used for rapid analyses because the same resolution can be generated in a shorter time.

7. The square root of resolution is proportional to the column i.d. The smaller the i.d., the greater will be the column efficiency and the shorter will be the time of analysis for a specific degree of resolution.

Film Thickness of the Stationary Phase

1. Retention and sample capacity increase with increasing film thickness with a concurrent decrease in column efficiency.

2. Film thickness is inversely proportional to plate number and almost directly proportional to the time of analysis.

3. Thin-film columns provide higher resolution of high-boiling solutes but lower resolution of more volatile components under any set of column temperature conditions.

4. The sample capacity of thin-film columns may be inadequate and require cryogenic temperature control of the column oven.

5. Film thicknesses below 0.2 μm permit the use of longer columns for complex samples.

6. A solute will exhibit a lower elution temperature as film thickness decreases; thus, thin-film columns are ideal for high-boiling petroleum fractions, triglycerides, and so on.

7. A thick-film column (which is inherently more inert) should be utilized for samples having a range of solute concentrations. Thicker films of stationary phase (> 1 μm) should be used for analysis of more volatile solutes. Very thick films (> 5 μm) should be selected for analyses to be performed at room temperature.

8. Thicker-film columns necessitate higher elution temperatures, but incomplete elution of all sample components may result.

9. Higher elution temperatures for prolonged periods of time result in a reduced column lifetime and more column bleed.

10. A capillary column 30 m or longer, with a thick film of stationary phase, offers an alternative to cryogenic oven temperature control for solute-focusing purposes, especially attractive with auxiliary sample introduction techniques of purge-and-trap and thermal desorption.

A summary of sample capacities for capillary columns of several inner diameters with different film thicknesses is given in Table 3.17.

Column Length. Resolution is a function of the square root of the number of theoretical plates or column length. One must consider the trade-off of the increase in overall resolution in a separation by augmenting column length with the simultaneous increase in analysis time under isothermal conditions. Prudence suggests using the shortest column length that will produce the necessary resolution. The sample capacity of a capillary column increases with column length. Increasing the length of a capillary column from 15 to 30 m, for example, results in an improvement by a factor of 1.4 (the square root of 2) in resolution, but analysis time also doubles [equation 3.15], which may limit sample throughput in a laboratory. To double resolution between two adjacent peaks, one needs a fourfold increase in column length. If one is already using a 30-m column, increasing the column length to

TABLE 3.17   Column Capacity as a Function of Inner Diameter and Film Thickness

Inner Diameter (mm)	Film Thickness (μm)	Capacity[a] (ng/component)
0.25	0.15	60–70
	0.25	100–150
	0.50	200–250
	1.0	350–400
0.32	0.25	150–200
	0.5	250–300
	1.0	400–450
	3.0	1200–1500
0.53	1.0	1000–1200
	1.5	1400–1600
	3.0	3000–3500
	5.0	5000–6000

Source: Data from the 1994–1995 J&W Scientific Catalog.

[a]Capacity is defined as the amount of component where peak asymmetry occurs at 10% at half-height.

120 m is unreasonable. Here a column of the same initial length (shorter in some cases) that has another stationary phase will have a different selectivity and solve the problem. The situation is slightly different under temperature-programming conditions, where a large improvement in resolution can sometimes be obtained with only a moderate increase in analysis time.

The best approach is the selection of a 25- or 30-m column for general analytical separations (Figure 3.40) and for fingerprinting chromatograms generated under the same chromatographic conditions for comparison of samples (Figure 3.41). A shorter length of column may be employed for rapid screening or simple mixtures or a 60-m column for very complex samples (a longer column also generates more column bleed). Temperature-programming ramp profiles can be adjusted to optimize resolution; computer software based on optimization of column temperature can be of great assistance in establishing favorable column temperature conditions (Chapter 4).

Capillary Columns of 0.53 mm i.d.: Megabore Columns

Many applications previously performed on a packed column can now be done with a *megabore column*, a capillary column of 0.53 mm i.d. Like a packed column, a megabore column with a fairly thick film of stationary phase has a low phase ratio and exhibits retention characteristics and sample capacities similar to those of a packed column. For example, the phase ratios of a 0.53-mm-i.d. column with film thicknesses of 3.00 and 5.00 μm are 44 and 26, respectively, and lie in the range of the phase ratio of a packed column. An interesting scanning electron micrograph of a cross section of a capillary column with a very thick film of stationary phase (ca. 18 μm) appears in Figure 3.42. Examination of the photograph indicates that the film thickness of the stationary phase is nearly identical to the thickness of the

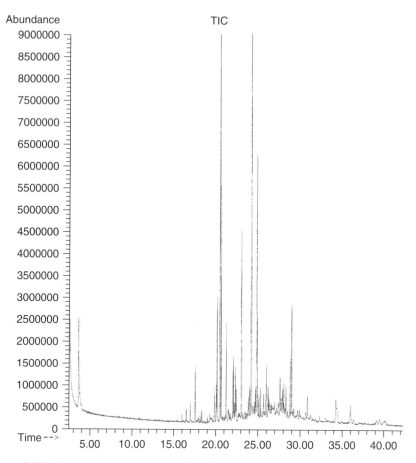

Figure 3.40   Total ion chromatogram of a chloroform extract of a wood sliver from a telephone pole for determination of pentachlorophenol. Conditions: 30 m × 0.25 mm i.d. DB-5 column with 0.25-μm film. Column conditions: 40°C at 8°C/min to 250°C after 1 min of isothermal hold; splitless injection of 1 μL (1 min delay time).

polyimide outer coating on this particular column; a thick film of stationary phase of this nature is a result of phase-height augmentation technology (PHAT).

A 0.53-mm-i.d. column offers the best of both worlds, because it combines the attributes of a fused-silica capillary column with the high sample capacity and ease of use of a packed column. For many applications, analytical methods developed with a packed column can easily be transferred to a megabore column with the appropriate stationary phase. Peaks generated with a megabore column typically are sharper and exhibit less tailing compared to those with a packed column. Redistribution of the stationary phase can occur at the inlet of a packed column with large injections of solvent and leave an exposure of silanol sites on a diatomaceous earth support. With a cross-linked phase in a megabore

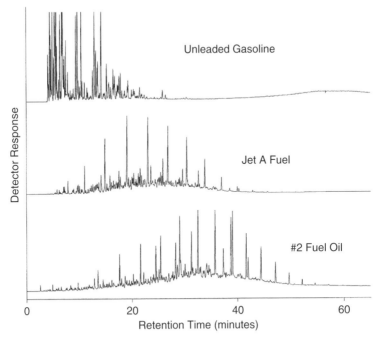

Figure 3.41 Chromatograms showing the separation of an unleaded gasoline, Jet A fuel, and No. 2 fuel oil under the same column conditions. Column 30 m × 0.25 mm i.d. HP-1 (0.25-μm film); temperature conditions 35°C (2 min) at 4°C/min to 260°C, detector FID, 25 cm/s He.

column, this problem is eliminated. Lewis acid sites, which are a problem with supports, are also absent in this larger-diameter capillary column. Analysis time is also shorter as a rule (Figure 3.43). On the other hand, long megabore columns can be used for the analysis of more complex samples, such as separation of the reference standard for EPA Method 502.2 (VOCs in drinking water) presented in Figure 3.44.

Correlation of Column Dimensions and Film Thickness with Parameters in the Fundamental Resolution Equation

Column length governs theoretical plates and time of analysis and has an associated cost factor. For isothermal separations, length affects retention times; changing the column length alters retention accordingly. When one considers increasing column length, one should also consider cost. In other words, one must weigh whether doubling the column length for a 41% increase in resolution justifies doubling the column cost and a twofold increase in analysis time. The situation is slightly different for analyses requiring temperature programming, where a twofold increase in column length increases analysis times only slightly and retention is much more dependent on column temperature.

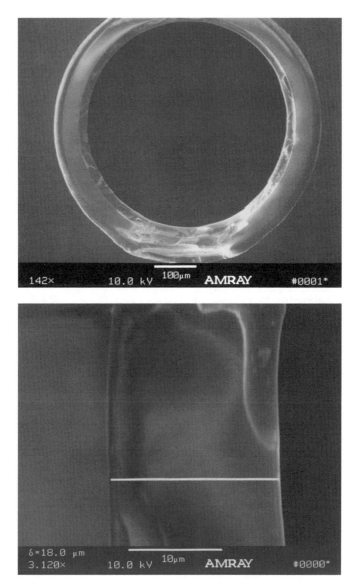

Figure 3.42 Scanning electron micrograph of the cross section of a 0.53-mm-i.d. column; the 18-μm stationary-phase film thickness is approximately equal to the thickness of the polyimide outer coating on the capillary. (Courtesy of the Quadrex Corporation.)

Column inner diameter affects not only resolution but also sample capacity. For example, the megabore column has an excellent capacity at the expense of a sacrifice in resolution offered by the narrower-diameter columns. Analysis times are also shorter, and of course, it remains the choice of choice for direct replacement of a packed column. But it is a column of choice for purity or trace component

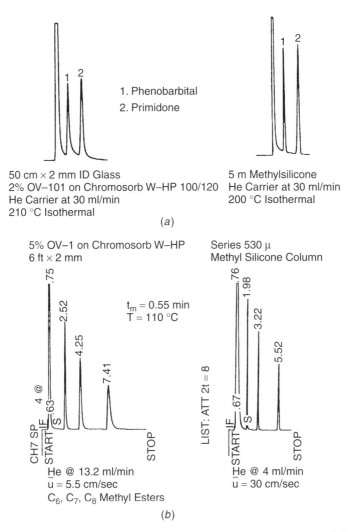

Figure 3.43 Chromatograms comparing the (*a*) effect of the inertness of a 0.53-mm-i.d. column to the more active surfaces within a packed column, and (*b*) the retention characteristics of a packed and 0.53-mm-i.d. column. (Courtesy of Agilent Technologies.)

analysis, large volume injection, and ancillary injection techniques. As column i.d. decreases, resolution increases at the expense of sample capacity and analysis time increases as well; the dynamic range of the detector may also be compromised. Column inner diameters of 0.10, 0.18, and 0.25 mm are recommended for fast GC.

Thick films are necessary for separations of low boilers, thin-film columns for high boilers. As the column i.d. increases, thicker films are often selected; thicker films of stationary phase also mean more column bleed (Section 3.8). One must

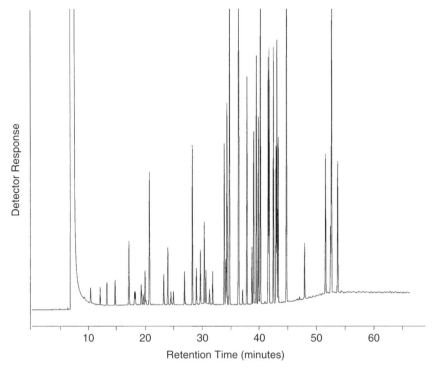

Figure 3.44   Chromatogram of the separation of reference standard for EPA Method 502.2 on a 75 m × 0.53 mm i.d. methyphenylcyanopropylsilicone capillary column (2.5-μm film). Column conditions: 35°C at 4°C/min to 280°C after 10 min of isothermal hold; 10 mL/min He; splitless injection and FID.

keep in mind the concept of phase ratio and its role in column selection. Two figures place the role of film thickness and i.d. into perspective: Figure 3.45 illustrates parallel separations on three columns that have the same length and phase ratio (250) but differ individually in i.d. and film thickness; Figure 3.45 depicts the impact that the numerator and denominator in the phase ratio expression can have on time of analysis.

Let us now review the fundamental resolution expression [equation (2.13)], which is repeated here for convenience, and examine each term on the right-hand side:

$$R_s = \frac{1}{4}N^{1/2}\frac{\alpha - 1}{\alpha}\frac{k}{1 + k} \tag{3.17}$$

Efficiency, N, is a function of the carrier gas (i.e., helium vs. hydrogen vs. nitrogen, and column length and diameter. Doubling column length yields a 41% in resolution; quadrupling of column length doubles resolution but at the expense of a fourfold increase in both time of analysis and cost. The question to be asked: Is it worth it?

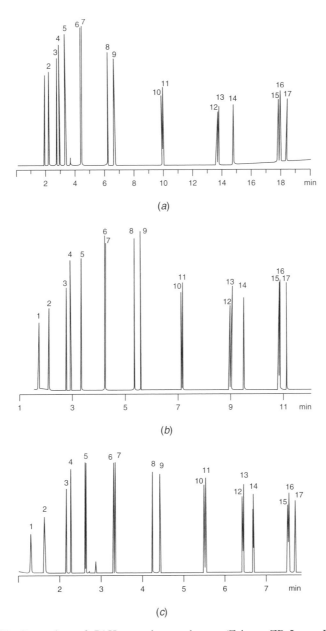

Figure 3.45 Separation of PAHs on three columns (Zebron ZB-5 ms, Phenomenex, Torrance, CA) of the same phase ratio of 250. *a*; 30 m × 0.25 mm, 0.25-μm film; *b* 20 m × 0.18 mm, 0.18-μm film *c* 10 m × 0.10 mm, 0.10-μm film. (1) Naphthalene, (2) 2-methylnaphthalene, (3) acenaphthalene, (4) acenaphthene, (5) fluorene, (6) phenanthrene, (7) anthracene, (8) fluoranthrene, (9) pyrene, (10) benzo[*a*]anthracene, (11) chrysene, (12) benzo[*b*]fluoranthrene, (13) benzo[*k*]fluoranthrene, (14) benzo[*a*]pyrene, (15) indeno[1,2,3-*cd*]pyrene, (16) dibenz[*a,h*]anthracene, (17) benz[*g,h,I*]perylene. Chromatograms courtesy of Phenomenex.

Selectivity, also known as the *band-spacing factor*, α, depends on column temperature and the nature of the stationary phase. Numerous examples of the effect of selectivity on a separation were presented earlier. Consider two separations: (1) where the α value of a critical band pair is 1.1, or $(\alpha - 1)/\alpha = 0.09$; and (2) where α of the same critical band pair is 1.4, or $(\alpha - 1)/\alpha = 0.29$. A change of 0.3 in α produces greater than a threefold enhancement in resolution. Prudence suggests that manipulation of column temperature should be the first approach investigated in improving resolution, then changing to a column of different polarity should be considered. Also note that resolution approaches zero as selectivity approaches unity (i.e., coelution).

The *retention factor*, k, is governed by the column temperature, the film thickness of the stationary phase, and the column diameter. The term $k/(1 + k)$ in equation (3.17) can not exceed unity and, in fact, can only asymptotically approach unity, as illustrated in Figure 3.46. Also note in Figure 3.46b that resolution increases rapidly until k reaches a value of approximately 10, after which there is no further contribution of this term to overall resolution, only an increase in the time of analysis, an important consideration in a laboratory of high sample throughput. This increase in time of analysis may or may not be necessary, depending on the number of components present in the sample.

Column Selection for Gas Chromatography by Specifications

Gas chromatographic column dimensions and an appropriate stationary phase are provided in Tables 3.18 to 3.21 for each designated ASTM, EPA, and NIOSH method and in Table 2.15 for USP methods. It is the goal of the authors to list in one place, as a convenient reference, tabulations of column dimensions associated with key analytical methods. The column cross-reference charts in Tables 3.13 and also provide additional column choices for any method. (The data in the following tables are also available at the Agilent Technologies and SGE Web sites.)

- USP (*United States Pharmacopoeia* and *National Formulary*). Column descriptions of the analysis of specific targeted analytes are provided in Table 2.15.
- ASTM (American Society for Testing and Materials). Suggested column configurations indicated in ASTM methods are listed in Table 3.18.
- EPA (Environmental Protection Agency). Table 3.19 is an outline of the EPA method numbers for the 500 and 600 series on drinking water, the series 8000 on solid waste, and the CLP series. Columns specific for each of these methods is detailed in Table 3.20.
- NIOSH (National Institute for Occupational Safety and Health). Capillary columns compatible for a wide variety of NIOSH methods are listed in Table 3.21.
- OSHA (Occupational Safety and Health Administration). The OSHA methods, listed alphabetically by both substance and method number, CAS number(s),

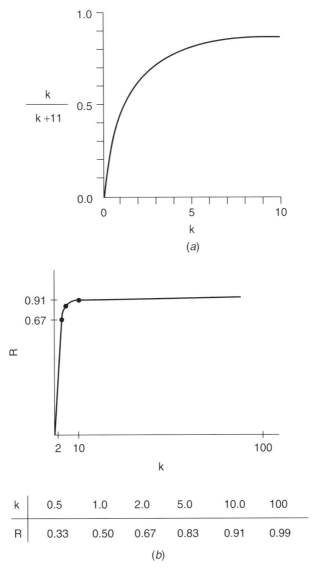

Figure 3.46   (*a*) Effect of retention factor term on *k* and (*b*) effect of *k* on resolution at constant *N* and selectivity.

type of instrumentation, column, and so on, are too extensive to appear in this chapter. However, the interested read is referred to the Web site of the U.S. Department of Labor, Occupational Safety and Health Administration, www.osha.gov/sampling and analytical methods/index of sampling and analytical methods.

TABLE 3.18 Suggested Capillary Column Dimensions for ASTM Methods

D1983	Standard Test Method for Fatty Acid Composition by Gas–Liquid Chromatography of Methyl Esters	30 m × 0.25 mm × 0.25 μm SolGel-WAX 30 m × 0.25 mm × 0.25 μm BP20
D2245	Oils and Oil Acids in Solvent-Reducible Paints	60 m × 0.25 mm × 0.25 μm BPX70
D2268	Standard Test Method for Analysis of High-Purity *n*-Heptane and Isooctane by Capillary GC	60 m × 0.25 mm × 0.50 μm BP1
D2306	Standard Test Method for C8 Aromatic Hydrocarbons by GC	60 m × 0.25 mm × 0.25 μm SolGel-WAX 60 m × 0.25 mm × 0.25 μm BP20
D2360	Standard Test Method for Trace Impurities in Monocylic Aromatic Hydrocarbons by GC	60 m × 0.32 mm × 0.25 μm SolGel-WAX
D2426	Standard Test Method for Butadiene Dimer and Styrene in Butadiene Concentrates by GC	30 m × 0.53 mm × 5.00 μm BP1
D2427	Standard Test Method for Determination of C2 through C5 Hydrocarbons in Gasoline by GC	30 m × 0.53 mm × 5.00 μm BP1
D2456	Polyhydric Alcohols in Alkyd Resin	30 m × 0.53 mm × 1.0 μm BP20
D2580	Phenols in Water	30 m × 0.32 mm × 0.50 μm BPX5
D2753	Oil and Oil Acids	60 m × 0.25 mm × 0.25 μm BPX70
D2800	FAME Analysis	60 m × 0.25 mm × 0.25 μm BPX70
D2804	Standard Test Method for Purity of Methyl Ethyl Ketone by GC	30 m × 0.53 mm × 1.00 μm BP20
D2887	Standard Test Method for Boiling Range Distribution of Petroleum Fractions by GC	10 m × 0.53 mm × 2.65 μm BPX1 5 m × 0.53 mm × 2.65 μm BPX1
D2998	Polyhydric Alcohols in Alkyd Resin	30 m × 0.32 mm × 1.0 μm BP1
D2999	Monopentaerythritol in Commercial Pentaerythritol	30 m × 0.53 mm × 1.0 μm BP1
D3009	Composition of Turpentine	30 m × 0.32 mm × 0.50 μm SolGel-WAX
D3054	Standard Test Method for Analysis of Cyclohexane by GC	60 m × 0.32 mm × 0.50 μm BP1
D3168	Polymers in Emulsion Paints	30 m × 0.32 mm × 1.0 μm BP1
D3257	Standard Test Method for Aromatics in Mineral Spirits by GC	30 m × 0.53 mm × 3.0 μm BP624
D3271	Solvent Analysis in Paints	30 m × 0.53 mm × 1.0 μm BP20
D3304	PCBs in Environmental Materials	60 m × 0.25 mm × 0.25 μm BPX5
D3328	Comparison of Waterborne Petroleum Oils	30 m × 0.32 mm × 3.0 μm BP1
D3329	Standard Test Method for Purity of Methyl Isobutyl Ketone by GC	30 m × 0.53 mm × 1.0 μm BP20 30 m × 0.53 mm × 3.0 μm BP624

TABLE 3.18 (*continued*)

D3432	Standard Test Method for Unreacted Toluene Diisocyanates in Urethane Prepolymers and Coating Solutions by GC	30 m × 0.32 mm × 1.0 μm BP1
D3447	Standard Test Method for Purity of Halogentated Organic Solvents	30 m × 0.53 mm × 3.0 μm BP624
D3452	Identification of Rubber	30 m × 0.53 mm × 1.0 μm BP1
D3457	FAME Analysis	60 m × 0.25 mm × 0.25 μm BPX70
D3534	Standard Test Method for PCBs in Water	30 m × 0.53 mm × 1.5 μm BPX5
D3545	Standard Test Method for Alcohol Content and Purity of Acetate Esters by GC	30 m × 0.53 mm × 3.0 μm BP624
D3687	Standard Test Method for Alcohol Content and Purity of Acetate Esters by GC	30 m × 0.53 mm × 1.0 μm BP20
D3695	Standard Test Method for Volatile Alcohols in Water by Direct Aqueous-Injection GC	30 m × 0.53 mm × 1.0 μm BP20
D3710	Boiling Range Distribution of Gasoline and Gasoline Fractions by GC	10 m × 0.53 mm × 2.65 μm BPX1 5 m × 0.53 mm × 2.65 μm BPX1
D3725	Fatty Acids in Drying Oils	30 m × 0.53 mm × 0.50 μm BPX70
D3760	Standard Test Method for Analysis of Isopropylbenzene (Cumene) by GC	60 m × 0.32 mm × 0.25 μm SolGel-WAX 60 m × 0.32 mm × 0.25 μm BP20 60 m × 0.32 mm × 0.50 μm BP1
D3797	Standard Test Method for Analysis of *o*-Xylene by GC	60 m × 0.32 mm × 0.50 μm SolGel-WAX
D3798	Standard Test Method for Analysis of *p*-Xylene by GC	60 m × 0.32 mm × 0.50 μm SolGel-WAX
D3871	Standard Test Method for Purgeable Organic Compounds in Water Using Headspace Sampling	60 m × 0.53 mm × 3.0 μm BP624
D3893	Standard Test Method for Purity of Methyl Amyl Ketone and Methyl Isoamyl Ketone by Gas Chromatography	30 m × 0.53 mm × 3.0 μm BP624
D3962	Impurities in Styrene	30 m × 0.53 mm × 1.0 μm BP21(FFAP)
D3973	Standard Test Method for Low-Molecular-Weight Halogenated Hydrocarbons in Water	30 m × 0.53 mm × 3.0 μm BP624
D4059	PCBs in Insulating Liquids	60 m × 0.25 mm × 0.25 μm BPX5
D4415	Standard Test Method for Determination of Dimer in Acrylic Acid	30 m × 0.32 mm × 0.25 μm SolGel-WAX

(*Continued*)

TABLE 3.18   (*continued*)

D4443	Standard Test Method for Residual Vinyl Chloride Monomer Content in PPB Range in Vinyl Chloride Homo- and Co-Polymers by Headspace GC	30 m × 0.53 mm × 3.0 μm BP624
D4735	Standard Test Method for Determination of Trace Thiophene in Refined Benzene by GC	30 m × 0.53 mm × 0.5 μm SolGel-WAX 30 m × 0.53 mm × 1.0 μm BP21
D4773	Standard Test Method for Propylene Gycol Monomethyl Ether, Dipropylene Glycol Monomethyl Ether, and Propylene Glycol Monmethyl Ether Acetate	30 m × 0.53 mm × 5.00 μm BP5
D4806	Denatured Fuel Ethanol for Blending with Gasolines for Use as Automotive Spark-Ignition Engine Fuel	100 m × 0.25 mm × 0.5 μm BP1 Uses Method D5501
D4815	Determination of MTBE, ETBE, TAME, DIPE, *tert*-Amyl Alcohol and C1 to C4 Alcohols in Gasoline by GC	30 m × 0.53 mm × 2.6 μm BP1
D4864	Determination of Traces of Methanol in Propylene Concentrates by GC	15 m × 0.53 mm × 1.5 μm BPX5 15 m × 0.53 mm × 1.5 μm BP5
D4947	Standard Test Method for Chlordane and Heptachlor Residues in Indoor Air	30 m × 0.53 mm × 1.5 μm BPX5 30 m × 0.53 mm × 1.0 μm BPX50
D5060	Standard Test Method for Determining Impurities in High-Purity Ethylbenzene by GC	60 m × 0.32 mm × 0.5 μm SolGel-WAX
D5075	Standard Test Method for Nicotine in Indoor Air	30 m × 0.32 mm × 1.0 μm BPX5
D5134	Standard Test Method for Detailed Analysis of Petroleum Naphthas Through *n*-Nonane by Capillary GC	60 m × 0.25 mm × 0.5 μm BP1
D5135	Standard Test Method for Analysis of Styrene by Capillary GC	60 m × 0.32 mm × 0.50 μm SolGel-WAX
D5399	Standard Test Method for Boiling Point Distribution of Hydrocarbon Solvents by GC	10 m × 0.53 mm × 2.65 μm BPX1
D5441	Analysis of Methyl *tert*-Butyl Ether by GC	60 m × 0.25 mm × 0.5 μm BP1 100 m × 0.25 mm × 0.5 μm BP1
D5442	Analysis of Petroleum Waxes by GC	25 m × 0.32 mm × 0.25 μm BP1 15 m × 0.25 mm × 0.25 μm BP5 15 m × 0.25 mm × 0.25 μm BPX5

TABLE 3.18 (*continued*)

D5480	Motor Oil Volatility by GC	12 m × 0.53 mm × 0.15 μm <u>HT5</u>
D5501	Determination of Ethanol Content of Denatured Fuel Ethanol by GC	100 m × 0.25 mm × 0.5 μm <u>BP1</u>
D5580	Determination of Benzene, Toluene, Ethylbenzene, *p/m*-Xylene, *o*-Xylene, C9 and Heavier Aromatics in Finished Gasoline by GC	30 m × 0.53 mm × 5.0 μm <u>BP1</u>
D5599	Determination of Oxygenates in Gasoline by GC and Oxygen Selective Flame Ionization Detection	60 m × 0.25 mm × 1.0 μm <u>BP1</u>
D5623	Sulfur Compounds in Light Petroleum Liquids by Gas Chromatography and Sulfur Selective Detection	30 m × 0.32 mm × 4.0 μm <u>BP1</u>
D5713	Standard Test Method for Analysis of High Purity Benzene for Cyclohexane Feedstock by Capillary GC	60 m × 0.25 mm × 0.5 μm <u>BP1</u>
D5739	Standard Practice for Oil Spill Source Identification by GC and Positive Ion Electron Impact Low Resolution Mass Spectrometry	30 m × 0.25 mm × 0.25 μm <u>BPX5</u>
D5769	Determination of Benzene, Toluene and Total Aromatics in Finished Gasolines by GC/MS	60 m × 0.32 mm × 5.0 μm <u>BP1</u> (cool on col) 60 m × 0.25 mm × 1.0 μm <u>BP1</u> (Splitter) 60 m × 0.32 mm × 1.0 μm <u>BP1</u> (Splitter)
D5790	Standard Test Method for Measurement of Purgeable Organic Compounds in Water by Capillary Column GC/MS	60 m × 0.25 mm × 1.4 μm <u>BP624</u>
D5812	Standard Test Method for Determination of Organochlotine Pesticides in Water by Capillary Column GC	30 m × 0.25 mm × 0.25 μm <u>BPX5</u> 30 m × 0.25 mm × 0.25 μm BP10(1701) 30 m × 0.25 mm × 0.25 μm <u>BPX50</u>
D5917	Standard Test Method for Trace Impurities in Monocyclic Aromatic Hydrocarbons by GC and External Calibration	60 m × 0.32 mm × 0.25 μm <u>SolGel-WAX</u>
D5974	Standard Test Methods for Fatty and Rosin Acids in Tall Oil Fraction Products by Capillary GC	60 m × 0.25 mm × 0.25 μm <u>BPX70</u>

(*Continued*)

TABLE 3.18   (*continued*)

D5986	Determination of Oxygenates, Benzene, Toluene, C8–C12 Aromatics and Total Aromatics in Finished Gasoline by GC/FTIR	60 m × 0.53 mm × 5.0 μm <u>BP1</u>
D6160	Standard Test Method for Determination of PCBs in Waste Materials by GC	30 m × 0.25 mm × 0.25 μm <u>BPX5</u>
D6352	Boiling Range Distribution of Petroleum Fractions by GC	10 m × 0.53 mm × 0.9 μm <u>BPX1</u> 5 m × 0.53 mm × 0.1 μm <u>BPX1</u>[a] 5 m × 0.53 mm × 0.17 μm <u>BPX1</u>[a]
D6417	Estimation of Engine Oil Volatility by Capillary GC	10 m × 0.53 mm × 0.9 μm <u>BPX1</u>
D6729	Test Method for Determination of Individual Components in Spark Ignition Engine Fuels by 100 Meter Capillary High Resolution Gas Chromatography	100 m × 0.25 mm × 0.5 μm <u>BP1</u>
D6730	Test Method for Determination of Individual Components in Spark Ignition Engine Fuels by 100 Meter Capillary (with Precolumn) High Resolution Gas Chromatography	100 m × 0.25 mm × 0.5 μm <u>BP1</u> 30 m × 0.25 mm × 1.0 μm <u>BP5</u> 15 m × 0.25 mm × 1.0 μm <u>BPX5</u>
E202	Standard Test Methods for Analysis of Ethylene Glycols and Propylene Glycols	30 m × 0.53 mm × 3.0 μm <u>BP624</u>
E475	Standard Test Method for Assay of Di-*tert*-Butyl Peroxide Using GC	30 m × 0.53 mm × 5.0 μm <u>BP5</u>
E1616	Standard Test for Analysis of Acetic Anhydride Using GC	50 m × 0.32 mm × 0.50 μm <u>BP1</u>
E1863	Standard Test Method for Analysis of Acrylonitrile by GC	60 m × 0.32 mm × 1.0 μm <u>BP20(Wax)</u>
Future Method (Under Review)		
	Boiling Range Distribution of Non-fully Eluting Petroleum Fractions by GC	10 m × 0.53 mm × 0.9 μm <u>BPX1</u>

Source: www.sge.com.

[a]Aluminum clad.

TABLE 3.19   EPA Method Number Designations for Three Sample Materials

Method	Method Title	Method	Method Title
EPA 500 Series Methods: Drinking Water		**EPA 8000 Series Methods: Solid Waste**	
502.2	Volatile Organic Compounds	8021	Volatile Organics
		8060	Phithalate Esters
504.1	1,2-Dibromoethane and 1,2-dibromo-3-chloropropane	8081	Chlorinated Pesticides
		8090	Nitroaromatics and Cyclic Ketones
508	Chlorinated Pesticides	8095	Explosives
515.1	Chlorophenoxyacid Herbicides	8100	Polynuclear Aromatic Hydrocarbons
524.1/.2	Volatile Organics	8140/8141/	Organophosphorus
525	Semivolatile Organics	8141A	Pesticides
526	Semivolatile Organics (screening)	8151	Chlorophenoxyacid Herbicides
528	Phenols	8240	Volatile Organics (8260 short list)
551.1	Chlorinated Disinfection By-products	8260B	Volatile Organics
552.2	Haloacetic Acids	8270	Semivolaltile Organics
		8330	Explosives by HPLC
		8095	Explosives by GC
EPA 600 Series Methods: Drinking Water		**CLP Methods**	
		OLC 03.2	Volatile Organics
601	Purgeable Halocarbons	CLP	Semivolatile Organics
602	Purgeable Aromatics	Appendix IX	Semivolatile Organics
604	Phenols	Skinner List	Semivolatile Organics
605	Benzidines		
606	Phthalate Esters		
607	Nitrosamines		
609	Nitroaromatics and Isophorone		
610	Polynuclear Aromatic Hydrocarbons (HPLC/GC)		
611	Haloethers		
612	Chlorinated Hydrocarbons		
615	Chlorophenoxyacid Herbidices		
619	Nitrogen/Phosphorus Herbicides and Pesticides		
625	Base/Neutrals, Acids, and Pesticides		

TABLE 3.20 Capillary Column Dimensions for EPA Methods

EPA Method Reference	Analyte Type	Sample Matrix	Common Sample Preparation	Detector Types	Column Suggested
Volatiles					
501	Trihalomethanes	Drinking water	Purge and trap, direct injection, headspace	ELCD, ECD	DB-VRX, 30 m × 0.45 mm × 2.55 µm
502.2, 8021, CLP-Volatiles	Volatile organic compounds (VOCs)	Drinking water, wastewater, solid wastes	Purge and trap, direct injection, headspace	PID, ELCD	DB-VRX, 75 m × 0.45 mm × 2.55 µm
601, 8010	Purgeable halogenated organics	Wastewater, solid wastes	Purge and trap, headspace for screening	PID, ELCD	DB-VRX, 75 m × 0.45 mm × 2.55 µm DB-624, 75 m × 0.45 mm × 2.55 µm
503.1, 602, 8020	Purgeable aromatic organics	Drinking water, trap, wastewater, solid wastes	Purge and headspace for screening	PID	DB-VRX, 30 m × 0.45 mm × 2.55 µm DB-VRX, 30 m × 0.45 mm × 2.55 µm
524.2, 624, 8240, 8260, CLP-VOCs	VOCs using MSD	Drinking water, wastewater, solid wastes	Purge and trap, direct injection, headspace	MSD	DB-VRX, 60 m × 0.25 mm × 1.4 µm DB-624, 60 m × 0.20 mm × 1.4 µm HP-VOC, 60 m × 0.20 mm × 1.1 µm

Method	Analyte	Matrix	Extraction	Detector	Column
524.2, 624, 8240, 8260, CLP-VOCs	VOCs using 5973 MSD	Drinking water, wastewater, solid wastes	Purge and trap direct injection, headspace	MSD (5973)	DB-VRX, 20 m × 0.18 mm × 1.0 µm DB-624, 20 m × 0.18 mm × 1.0 µm
504.1, 8011	EDB and DBCP	Drinking water, solid wastes	Microextraction with hexane	ECD	DB-VRX, 30 m × 0.45 mm × 2.55 µm DB-624, 30 m × 0.45 mm × 2.55 µm
603, 8015, 8031	Acrylonitrile and acrolein	Wastewater, solid wastes	Purge and trap, liquid extraction, sonication	FID, NPD	DB-VRX, 30 m × 0.45 mm × 2.55 µm
Semivolatiles					
525, 625, 8270	Semivolatile organic compounds	Drinking water, wastewater, solid wastes	Liquid extraction, sonication, Soxhlet extraction, SPE	MSD	HP-5 ms, 30 m × 0.25 mm0.5 µm DB-5.625, 20 m × 0.18 mm × 0.36 µm
528, 604, 8040, 8041	Phenois	Wastewater, solid wastes	Liquid extraction, sonication, Soxhlet extraction, derivatization	ECD, RID	DB-5 ms, 30 m × 0.25 mm × 0.25 µm DB-XLB, 30 m × 0.25 mm × 0.25 µm DB-5 ms, 30 m × 0.53 mm × 1.5 µm DB-608, 30 m × 0.53 mm × 0.5 µm
506, 606, 8060, 8061	Phthalate esters	Drinking water, wastewater, solid wastes	Liquid extraction, sonication, Soxhlet extraction, SPE	ECD, FID	DB-5 ms, 30 m × 0.25 mm × 0.25 µm DB-5 ms, 30 m × 0.53 mm × 1.5 µm DB-608, 30 m × 0.53 mm × 0.5 µm

(Continued)

TABLE 3.20 (*continued*)

EPA Method Reference	Analyte Type	Sample Matrix	Common Sample Preparation	Detector Types	Column Suggested
605	Benzidines	Wastewater	Liquid extraction	ECD	DB-5 ms, 30 m × 0.25 mm × 0.25 μm DB-5 ms, 30 m × 0.53 mm × 1.5 μm DB-608, 30 m × 0.53 mm × 0.5 μm
607, 8070	Nitrosamines	Wastewater, solid wastes	Liquid extraction, sonication, Soxhlet extraction, SPE	NPD	DB-5 ms, 30 m × 0.25 mm × 0.25 μm DB-5 ms, 30 m × 0.25 mm × 1.5 μm
609, 8090	Nitroaromatics and isophorone	Wastewater, solid wastes	Soxhlet extraction, SPE	ECD, FID	HP-5 ms, 30 m × 0.25 mm × 0.5 μm DB-5 ms, 30 m × 0.53 mm × 1.5 μm DB-608, 30 m × 0.53 mm × 0.5 μm
610, 8100	Polynuclear aromatic hydrocarbons (PAHs)	Wastewater, solid wastes	Liquid extraction, sonication, Soxhlet extraction, SPE	FID	DB-5 ms, 30 m × 0.25 mm × 0.25 μm DB-5 ms. 30 m × 0.32 mm × 0.25 μm DB-1 ms, 30 m × 0.25 mm × 0.25 μm
612, 8120, 8121	Chlorinated hydrocarbons	Wastewater, solid wastes	Liquid extraction, sonication, Soxhlet extraction, SPE	ECD	DB-5 ms, 30 m × 0.32 mm × 0.5 μm HP-5 ms, 30 m × 0.32 mm × 0.5 μm DB-1, 30 m × 0.25 mm × 0.5 μm

Method	Compounds	Matrix	Sample Preparation	Detector	Column
551, 551.1A	Chlorinated disinfection by-products	Drinking water	Liquid extraction, derivatization	ECD	DB-5 ms, 30 m × 0.25 mm × 1.0 μm DB-5 ms, 30 m × 0.25 mm × 1.0 μm DB-1, 30 m × 0.25 mm × 1.0 μm
552, 552.1, 552.2	Halogenated acetic acids	Drinking water	Liquid extraction, derivatization	ECD	DB-35 ms, 30 m × 0.32 mm × 0.25 μm DB-XLB, 30 m × 0.32 mm × 0.5 μm

Pesticides, Herbicides, and PCBs

Method	Compounds	Matrix	Sample Preparation	Detector	Column
508.1, 608, 8081A, 8082, CLP-Pesticides	Organochlorine pesticides and PCBs	Drinking water, wastewater, solid wastes	Liquid extraction, sonication, Soxhlet extraction, SPE	ECD	DB-35 ms, 30 m × 0.32 mm × 0.25 μm DB-XLB, 30 m × 0.32 mm × 0.5 μm
515, 615, 8150, 8151	Phenoxy acid herbicides	Drinking water, wastewater, solid wastes	Liquid extraction, sonication, Soxhlet extraction, SPE	ECD	DB-35 ms, 30 m × 0.32 mm × 0.25 μm DB-XLB, 30 m × 0.32 mm × 0.5 μm
507, 614, 619, 622, 8140, 8141A	N- and P-containing pesticides and herbicides	Drinking water, wastewater, solid wastes	Liquid extraction, sonication, Soxhlet extraction, SPE	NPD, ELCD, FPD, MSD	DB-35 ms, 30 m × 0.25 mm × 0.25 μm DB-5 ms, 30 m × 0.25 mm × 0.25 μm
	PCB congeners using MSD			MSD	DB-XLB, 30 m × 0.18 mm × 0.18 μm DB-XLB, 60 m × 0.25 mm × 0.25 μm

Source: www.chem.agilent.com.

TABLE 3.21   Suggested Capillary Column Dimensions for NIOSH Methods

NIOSH Method	Method Name	Agilent J&W Phase Recommendation
1000	Allyl chloride	DB-VRX 30 m × 0.25 mm × 1.4 μm
1001	Methyl chloride	DB-VRX 30 m × 0.25 mm × 1.4 μm
1002	Chloroprene	DB-VRX 30 m × 0.25 mm × 1.4 μm
1003	Halogenated hydrocarbons	DB-VRX 30 m × 0.25 mm × 1.4 μm
1004	*sym*-Dichloroethyl ether	DB-VRX 30 m × 0.25 mm × 1.4 μm
1005	Methylene chloride	DB-VRX 30 m × 0.25 mm × 1.4 μm
1006	Trichlorofluoromethane	DB-VRX 30 m × 0.25 mm × 1.4 μm
1007	Vinyl chloride	DB-VRX 30 m × 0.25 mm × 1.4 μm
1008	Ethylene dibromide	DB-VRX 30 m × 0.25 mm × 1.4 μm
1009	Vinyl bromide	DB-VRX 30 m × 0.25 mm × 1.4 μm
1010	Epichlorohydrin	DB-WAX 15 m × 0.32 mm × 0.5 μm
1011	Ethyl bromide	DB-VRX 30 m × 0.25 mm × 1.4 μm
1012	Dibromodifluoromethane	DB-VRX 30 m × 0.25 mm × 1.4 μm
1013	1,2-Dichloropropane	DB-VRX 30 m × 0.25 mm × 1.4 μm
1014	Methyl iodide	DB-VRX 30 m × 0.25 mm × 1.4 μm
1015	Vinylidine chloride	DB-VRX 30 m × 0.25 mm × 1.4 μm
1016	1,1,1,2-Tetrachloro-2,2-difluoroethane and 1,1,2,2-tetrachloro-1,2-difluoroethane	DB-VRX 30 m × 0.25 mm × 1.4 μm
1017	Bromotrifluoromethane	DB-VRX 30 m × 0.25 mm × 1.4 μm
1018	Dichlorodifluoromethane and 1,2-dichlorotetrafluoroethane	DB-VRX 30 m × 0.25 mm × 1.4 μm
1019	1,1,2,2-Tetrachloroethane	DB-VRX 30 m × 0.25 mm × 1.4 μm
1020	1,1,2-Trichloro-1,2,2-trifluoroethane	DB-VRX 30 m × 0.25 mm × 1.4 μm
1022	Trichloroethylene	DB-VRX 30 m × 0.25 mm × 1.4 μm
1024	1,3-Butadiene	GS-Alumina 30 m × 0.53 mm
1300	Ketones 1	DB-WAX 30 m × 0.32 mm × 0.5 μm
1301	Ketones 2	DB-WAX 30 m × 0.32 mm × 0.5 μm
1400	Alcohols 1	DB-WAX 30 m × 0.32 mm × 0.5 μm
1401	Alcohols 2	DB-WAX 30 m × 0.32 mm × 0.5 μm
1402	Alcohols 3	DB-WAX 30 m × 0.32 mm × 0.5 μm
1403	Alcohols 4	DB-WAX 15 m × 0.32 mm × 0.5 μm
1450	Esters 1	DB-WAX 30 m × 0.32 mm × 0.5 μm
1500	Hydrocarbons, BP 36-126uC	DB-1 30 m × 0.25 mm × 0.25 μm
1501	Hydrocarbons, aromatic	DB-5 ms 30 m × 0.25 mm × 0.25 μm
1550	Naphthas	DB-1 60 m × 0.25 mm × 0.25 μm
1551	Turpentine	DB-1 60 m × 0.25 mm × 0.25 μm
1600	Carbon disulfide	GS-Q 30 m × 0.53 mm
1602	Dioxane	DB-WAX 15 m × 0.32 mm × 0.5 μm
1603	Acetic acid	DB-FFAP 15 m × 0.25 mm × 0.25 μm
1604	Acrylonitrile	DB-WAX 15 m × 0.32 mm × 0.5 μm
1606	Acetonitrile	DB-WAX 15 m × 0.32 mm × 0.5 μm
1608	Glycidol	DB-WAX 15 m × 0.32 mm × 0.5 μm
1609	Tetrahydrofuran	DB-1 15 m × 0.25 mm × 0.25 μm
1610	Ethyl ether	DB-WAX 15 m × 0.32 mm × 0.5 μm
1611	Methylal	DB-WAX 15 m × 0.32 mm × 0.5 μm
1612	Propylene oxide	DB-WAX 15 m × 0.32 mm × 0.5 μm

TABLE 3.21 (*continued*)

NIOSH Method	Method Name	Agilent J&W Phase Recommendation
1613	Pyridine	DB-WAX 15 m × 0.32 mm × 0.5 μm
1614	Ethylene oxide	DB-WAX 15 m × 0.32 mm × 0.5 μm
1615	Methyl-*tert*-butyl ether	DB-WAX 15 m × 0.32 mm × 0.5 μm
2000	Methanol	DB-WAX 15 m × 0.32 mm × 0.5 μm
2001	Cresol, all isomers	DB-WAX 30 m × 0.25 mm × 0.25 μm
2002	Amines, aromatic	DB-5 ms 30 m × 0.25 mm × 1.0 μm
2003	1,1,2,2-Tetrabromoethane	DB-VRX 30 m × 0.25 mm × 1.4 μm
2004	Dimethylacetamide and dimethylformamide	DB-WAX 15 m × 0.32 mm × 0.5 μm
2005	Nitrobenzenes	DB-WAX 15 m × 0.32 mm × 0.5 μm
2007	Aminoethanol compounds	DB-1 15 m × 0.25 mm × 1.0 μm
2010	Amines, aliphatic	CAM 15 m × 0.32 mm × 0.25 μm
2500	2-Butanone	DB-WAX 15 m × 0.32 mm × 0.5 μm
2501	Acrolein	DB-WAX 15 m × 0.32 mm × 0.5 μm
2503	Mevinphos	DB-5 ms 15 m × 0.25 mm × 0.25 μm
2504	Tetraethyl pyrophosphate	DB-1 15 m × 0.25 mm × 0.25 μm
2505	Furfuryl alcohol	DB-WAX 15 m × 0.32 mm × 0.5 μm
2506	Acetone cyanohydrin	DB-1 15 m × 0.25 mm × 1.0 μm
2507	Nitroglycerine and ethylene glycol dinitrate	DB-WAX 15 m × 0.32 mm × 0.5 μm
2508	Isophorone	DB-1 15 m × 0.25 mm × 0.25 μm
2510	1-Octanethiol	DB-5 ms 15 m × 0.25 mm × 0.25 μm
2513	Ethylene chlorohydrin	DB-WAX 15 m × 0.32 mm × 0.5 μm
2515	Diazomethane	DB-1 15 m × 0.32 mm × 0.25 μm
2516	Dichlorofluoromethane	DB-VRX 30 m × 0.25 mm × 1.4 μm
2517	Pentachloroethane	DB-5 ms 30 m × 0.25 mm × 0.5 μm
2518	Hexachloro-1,3-cyclopentadiene	DB-VRX 30 m × 0.25 mm × 1.4 μm
2519	Ethyl chloride	DB-VRX 30 m × 0.32 mm × 1.8 μm
2520	Methyl bromide	DB-VRX 30 m × 0.32 mm × 1.8 μm
2521	Methylcyclohexanone	DB-WAX 30 m × 0.32 mm × 0.5 μm
2522	Nitrosamines	DB-5 ms 30 m × 0.25 mm × 0.5 μm
2523	1,3-Cyclopentadiene	DB-1 15 m × 0.32 mm × 1.0 μm
2524	Dimethylsulfate	DB-WAX 15 m × 0.32 mm × 0.5 μm
2525	1-Butanethiol	DB-1 15 m × 0.32 mm × 1.0 μm
2526	Nitroethane	DB-WAX 15 m × 0.32 mm × 0.5 μm
2527	Nitromethane	DB-5 ms 30 m × 0.25 mm × 0.25 μm
2528	2-Nitropropane	DB-5 ms 30 m × 0.25 mm × 0.25 μm
2529	Furural	DB-WAX 30 m × 0.32 mm × 0.5 μm
2530	Biphenyl	DB-5 ms 15 m × 0.25 mm × 0.25 μm
2531	Gluteraldehyde	DB-WAX 30 m × 0.32 mm × 0.5 μm
2533	Tetraethyllead (as Pb)	DB-1 15 m × 0.25 mm × 0.25 μm
2534	Tetramethyllead (as Pb)	DB-1 15 m × 0.25 mm × 0.25 μm
2536	Valeraldehyde	DB-WAX 15 m × 0.32 mm × 0.5 μm
2537	Methyl methacrylate	DB-WAX 15 m × 0.32 mm × 0.5 μm
2538	Acetaldehyde	DB-1301 15 m × 0.32 mm × 1.0 μm
2539	Aldehydes, screening	DB-1 30 m × 0.32 mm × 0.25 μm
2541	Formaldehyde	DB-1701 30 m × 0.25 mm × 0.25 μm
3502	Phenol	DB-5 ms 15 m × 0.25 mm × 0.25 μm
3700	Benzene	DB-WAX 15 m × 0.32 mm × 0.5 μm

TABLE 3.21   *(continued)*

NIOSH Method	Method Name	Agilent J&W Phase Recommendation
3702	Ethylene oxide	DB-WAX 30 m × 0.32 mm × 0.5 μm
4000	Toluene	DB-5 30 m × 0.25 mm × 0.25 μm
5012	EPN, malathion, and parathion	DB-5 ms 15 m × 0.25 mm × 0.25 μm
5014	Chlorinated terphenyl (60% chlorine)	DB-5 ms 15 m × 0.25 mm × 0.25 μm
5017	Dibutyl phosphate	DB-5 ms 15 m × 0.25 mm × 0.25 μm
5019	Azelaic acid	DB-1 15 m × 0.32 mm × 0.25 μm
5020	Dibutyl phthalate and di(2-ethylhexyl)phthalate	DB-5 ms 15 m × 0.25 mm × 0.25 μm
5021	*o*-Terphenyl	DB-1 30 m × 0.25 mm × 0.25 μm
5025	Chlorinated diphenyl ether	DB-5 ms 15 m × 0.25 mm × 0.25 μm
5029	4,4-Dimethylenedianiline	DB-5 15 m × 0.25 mm × 0.25 μm
5500	Ethylene glycol	DB-WAX 15 m × 0.32 mm × 0.5 μm
5502	Aldrin and lindane	DB-5 ms 15 m × 0.25 mm × 0.25 μm
5503	Polychlorobiphenyls	DB-5 ms 30 m × 0.25 mm × 0.25 μm
5506	Polynuclear aromatic hydrocarbons	DB-5 ms 30 m × 0.25 mm × 0.25 μm
5509	Benzidine and 3,3-dichlorobenzidine	DB-5 15 m × 0.53 mm × 1.5 μm
5510	Chlordane	DB-5 ms 15 m × 0.25 mm × 0.25 μm
5514	Demeton	DB-5 15 m × 0.25 mm × 0.25 μm
5515	Polynuclear aromatic hydrocarbons (in the presence of isocyanates)	DB-5 ms 30 m × 0.25 mm × 0.25 μm
5516	2,4- and 2,6-Toluenediamine	DB-5 30 m × 0.25 mm × 0.25 μm
5517	Polychlorobenzenes	DB-1 15 m × 0.25 mm × 0.25 μm
5518	Naphthylamines	DB-5 ms 30 m × 0.25 mm × 0.25 μm
5519	Endrin	DB-5 ms 30 m × 0.25 mm × 0.25 μm

Source: www.chem.agilent.com.

3.8   COLUMN INSTALLATION AND CARE

Carrier Gas Purifiers

The discussion of carrier gas purity pertains to detector gases as well. Most labs traditionally have used compressed gas cylinders, but today, primarily for safety and practicality, gas generators are becoming increasingly common. If you choose to use cylinders, one needs to be aware of gas purity; otherwise, the detrimental presence of water and oxygen in a carrier gas stream can cause premature column fatigue due to degradation of the stationary phase in many cases while hydrocarbons can contribute to column bleed levels and FID detector noise over and above the contribution from detector gases. If the carrier gas is not research-grade or highest purity, it is always wise to employ moisture, oxygen, and hydrocarbon traps installed in a carrier gas line prior to entrance into a gas chromatograph, whether the gas flow is supplied by a single cylinder or by the line connected to a manifold located external to the laboratory. Column vendors offer arrays of traps for every gas-delivery application.

Much like the debate over the correct purity for chromatographic gases, analysts have debated the use of in-line gas purifiers versus ultrahigh-purity gases. Because there are many sources of contaminants in addition to the gas cylinder, we recommend using gas purifiers to protect your instruments. Often, the greatest sources of contaminants are introduced during the process of changing cylinders, which creates an opportunity for room air to enter both line and cylinder. In-line purifiers remove this surge of impurities and keep them from entering the instrument. The second source of contaminants is the diaphragm in the regulator. Most butyl rubber diaphragms will emit hydrocarbons into the gas stream, also contributing to column bleed. It is best to use regulators with stainless steel diaphragms with all carrier gas lines.

Gas generators such as those illustrated in Figure 3.47 can greatly simplify plumbing systems and eliminate the need for handling high-pressure and flammable materials. Because typically, these compact units can be located very near the instruments they serve, they eliminate the need for long gas lines and cylinders mounted in hallways. Compact, high-purity, worry-free and safe generators of nitrogen, air, and hydrogen are available. Hydrogen generators, in particular, provide important safety advantages. Relative to cylinders, the total amount of stored gas is small and pressures are much lower. This reduces the risk of explosion significantly. Safety devices internal to most generators shut down the units when the pressure surges or suddenly drops. Maintenance time spent on generators is less than that spent on changing cylinders.

There are a few negative aspects to the use of gas generators. They do require occasional maintenance. Hydrogen generators have many built-in safety devices. Nitrogen and air generators depend on house air systems or stand-alone air compressors. House air systems do go down occasionally and are seldom monitored for their moisture and hydrocarbon content. Bartram's comprehensive review of gas management systems for gas chromatography (135) is strongly recommended reading for all practicing gas chromatographers.

Ferrule Materials and Fittings

Ferrules for capillary columns are usually fabricated from graphite and Vespel–graphite composites. Graphite ferrules are easy to use, have a higher temperature limit, but are softer, more easily deformed, and are not recommended for GC-MS systems. Vespel–Graphite composite ferrules are harder and thus do not deform as easily and are therefore recommended for GC-MS systems. The characteristics of these materials are presented in Table 2.14. An alternative ferrule technology has been emerged with the SilTite metal ferrules. Graphite and graphite–vespel composites are made from different materials to the metal nuts, so the connection components expand and contact at different rates as the column oven temperature changes. With a metal ferrule system, since the components are made from the same material, the components expand and contract at the same

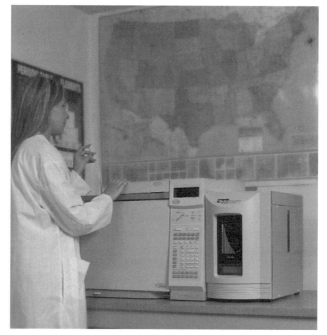

(a)

(b)

Figure 3.47 Selected gas generators: (a) FID gas station that delivers both high-purity hydrogen and air; (b) hydrogen generator; (c) nitrogen generator. (Photographs courtesy of Parker Balston Analytical Gas Systems.)

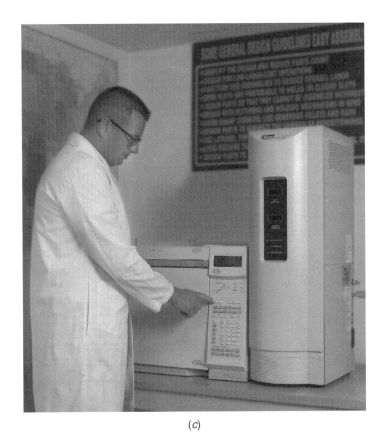

(*c*)

Figure 3.47   (*continued*)

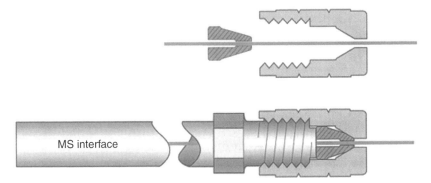

MS interface

Figure 3.48   All-metal SilTite connection. (Courtesy of SGE Analytical Sciences.)

rate during changes in oven temperatures. SilTite metal ferrules have an inherently higher temperature limit, and being metallic, the risk of MS contamination is eliminated. This connection arrangement is shown in Figure 3.48. However, because of the ubiquitous presence of graphite, Vespel, and Vespel–graphite composites as ferrule materials in capillary GC, some additional properties are summarized as follows:

- *Graphite* (100%): 450°C upper temperature limit; general purpose for capillary columns; suitable for FID, NPD, and ECD; recommended for high-temperature and cool on-column applications; material is free of sulfur and other contaminants; can be reused or removed easily; proper installation requires a finger-tight turn on the nut, then an additional quarter-turn with a wrench (as is the case with most fitting connections); *not for use with MS or oxygen-sensitive detectors.*

- *Graphite–Vespel* (15%:85%): 350°C upper temperature limit; general-purpose use in capillary columns; *recommended for MS or oxygen-sensitive detectors*; also compatible with other detectors, such as FID and NPD; most reliable leak-free connection; proper installation requires a finger-tight turn on the nut, then an additional quarter-turn with a wrench; the ferrule hole *must* match the column o.d. exactly to ensure a leak-free seal; not reusable.

- *Vespel* (100%): 280°C upper temperature limit; isothermal operation; can be reused or removed easily; excellent sealing material when making metal or glass connection; long lifetime; leaks after temperature cycle.

It is important to select the proper ferrule i.d. to be compatible with the o.d. of the capillary column, or a carrier gas leak will result after installation. Ferrules with an i.d. of 0.4 mm are recommended for 0.25-mm-i.d. columns, 0.5-mm-i.d. ferrules are recommended for 0.32-mm-i.d. columns, and 0.8-mm-i.d. ferrules are recommended for 0.53-mm-i.d. capillary columns.

Outlined below are guidelines for the preparation of a capillary column for installation.

1. Slide the retaining fitting or nut over the end of a new column, then the ferrule, and position them at least 6 in. away from the column end. It will be necessary to cut several inches from the end of the column because ferrule particles may have entered the column and can cause tailing and adverse adsorptive effects.

2. With a scoring tool, gently scribe the surface of the column several inches away from the end. While holding the column on each side of the scoring point, break the end at the scoring point at a slight downward angle. Any loose chips of fused silica or polyimide will fall away and will not enter the column, as may happen if it is broken in a completely horizontal configuration. This procedure eliminates the possibility of chips of fused silica or polyimide from

Figure 3.49   Termination of an end of a fused-silica capillary column with a ceramic wafer.

residing in the end of the column. Alternatively, an excellent cut can be made with a ceramic wafer (Figure 3.49).

3. Examine the end of the column with a 10× to 20× magnifier or an inexpensive light microscope. The importance of a properly made cut cannot be overstated. As illustrated in Figure 3.50, where a series of problematic scenarios are clearly evident, an improperly cut column can generate active sites and may cause peak tailing, peak splitting, or solute adsorption.

4. A supply of spare ferrules and related tools (Figures 3.51 and 3.52) are convenient to have in the laboratory when removal or changing columns is required. Ferrules, ferrule-handling accessories, tool kits, and other handy gadgets are available from many column manufacturers.

Column Installation

Define the injector and detector ends of the column. Align the end to be inserted into the injector with a ruler and mark the recommended distance of insertion as specified in the instrument manual with typewriter correction fluid. Then slide the ferrule and nut closer to the point of application of the correction fluid and mount the column cage on the hanger in the column oven. Alleviate any stress, sharp bends, or contact with sharp objects along the ends of the column. Insert the measured end into the injector and tighten the fitting. If the column is to be conditioned, leave the detector end of the column disconnected; otherwise, insert this end into the jet tip of the FID at the specified recommended distance, usually 2 mm down from the top of the jet. A quick-connect fitting is shown in Figure 3.53. This device facilitates column installation in most gas chromatographs without the use of wrenches and extends the ferrule lifetime.

Column Conditioning

Conditioning of a capillary column removes residual volatiles from the column. There are three essential rules for conditioning a capillary column, the first two of which also apply to the conditioning of a packed column:

1. Carrier gas flow must be maintained at all times when the column temperature is above ambient temperature, and there should be no gas leaks.

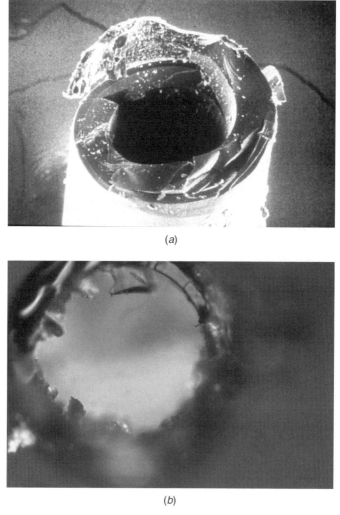

(a)

(b)

Figure 3.50 Scenarios that can occur in cutting an end of a capillary column. (a) Jagged protrusion; (b) jagged inner diameter; (c) particles of fused silica accumulated at inlet; (d) piece of polyimide attached to column end; (e) an acceptable cut. (Courtesy of Agilent Technologies.)

2. Do not exceed the maximum allowable temperature limit of the stationary phase or the column may be damaged permanently.

3. As opposed to the conditioning of a packed column, overnight conditioning of a capillary column is usually unnecessary. Instead, purge the column with normal carrier gas flow for 30 min at room temperature, then temperature program the column oven at 4°C/min to a temperature 20°C above the anticipated highest temperature at which the column will be subjected without exceeding the maximum allowable temperature limit. Usually, after a column has been maintained at this elevated temperature for several hours, a steady baseline is obtained and the column is ready for analyses to be conducted.

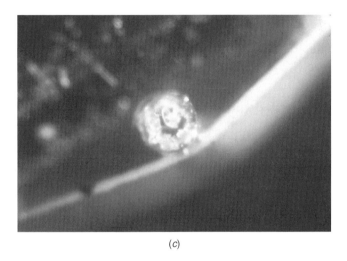

(c)

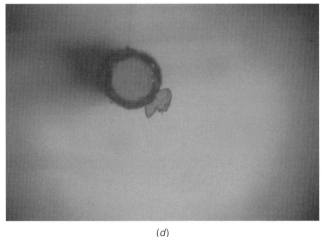

(d)

Figure 3.50 (*continued*)

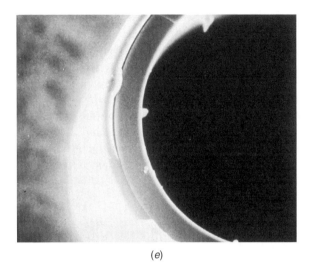

(e)

Figure 3.50   (*continued*)

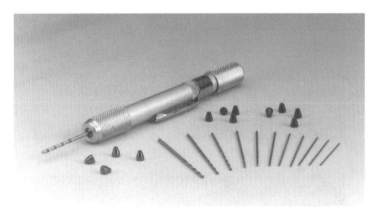

Figure 3.51   Capillary column ferrule kit containing an assortment of ferrules and a pin vise drill with bits for drilling or enlarging the bore of ferrules. (Reprinted with permission of Sigma-Aldrich Co.)

Use of high-purity carrier gas, a leak-free chromatographic system, and following the outline in reference 135 will greatly extend the lifetime of any column, packed or capillary. Other details pertaining to conditioning and column care are given in Section 2.4.

Column Bleed

Column bleed is a term used to describe the rise in baseline during a blank temperature programming run and is the inevitable consequence of increasing vapor

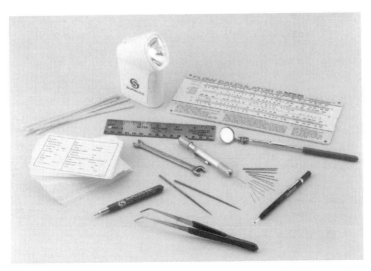

Figure 3.52 Capillary column tool kit containing tweezers, needle files, scoring tool, pin vise drill kit, flow calculator, pocket mirror, mini-flashlight, labels, septum puller, stainless steel ruler, and pipe cleaners. (Reprinted with permission of Sigma-Aldrich Co.)

pressure and thermal degradation of a polymer with an increase in column temperature as well as due to the accumulation of nonvolatiles in the column, as shown in Figure 3.54. One should expect some degree of bleeding with every column; some phases just generate more bleed than others. The detrimental effect of bleed due to column fatigue on a separation is illustrated in Figure 3.54c. Always try to select a stationary phase of high thermal stabilty. For example, a nonpolar phase bleeds less than a polar phase because the former typically has a higher temperature limit and thus is more thermally stable. Moreover, in comparing capillary columns of different dimensions, the level of bleed will increase with increasing amount of stationary phase in the column. Therefore, longer and wider-diameter columns yield more bleed than do shorter or narrower columns. Similarly, column bleed increases with increasing film thickness of the stationary phase and with increasing column length. Bleed is important during low-concentration GC-MS analyses and when it affects quantitation directly (i.e., when the bleed and solutes produce the same ions). In Figure 3.55, the structures of pertinent cyclic species associated with the breakdown of polysiloxane stationary phases and corresponding mass ions are presented. The process responsible for the formation of these fragments is often referred to as "back-biting." Reducing film thickness and using a shorter or narrower column will result in less bleeding.

The rate of temperature programming or ramp rate can influence the bleed profile from a column. As the rate of temperature programming increases, column bleed also increases. Finally, the more sensitive element-specific detectors (e.g., an ECD or NPD) will generate a more pronounced bleed profile if the stationary phase contains a heteroatom or functional group (–CN or –F) to which a detector responds in a sensitive fashion.

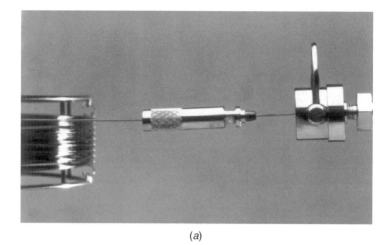

(a)

(b)

Figure 3.53   Quick-connect fitting for the installation of capillary columns; (*a*) internal sealing mechanism and (*b*) sealing mechanism inserted into the fitting and locked in place. (Courtesy of the Quadrex Corporation.)

Retention Gap and Guard Columns

A 0.5- to 5.0-m length of deactivated fused-silica tubing installed between the injector and the analytical column is often referred to as a *retention gap* or *guard column* (Figure 3.56) The term *retention gap* is used to describe this segment for on-column injection where the condensed solvent resides after injection but both solvent and solutes are not retained once vaporization occurs via temperature programming. As a guard column, this short length of deactivated tubing preserves the lifetime of an analytical column by collecting nonvolatile components and particulate matter in dirty samples that would otherwise accumulate at the inlet of the analytical column. As such, its latter role in capillary GC parallels the function of the guard column in HPLC. A guard column is considered to be a consumable item, requiring replacement from time to time, usually when the detector response of active analytes begins to decrease substantially. It eliminates the need for the repetitive removal of small sections at the inlet end of an analytical column with the buildup of contamination.

Proper implementation of the connection between the guard and the analytical columns is essential for the preservation of the chromatographic integrity of the system. The generation of active sites within the fitting can cause adsorptive losses and peak tailing. Commercially available fittings for this purpose include a metal butt connector of low dead volume, press-tight connectors, and a capillary

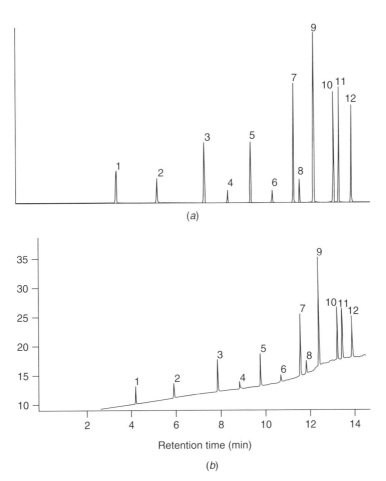

Figure 3.54 Comparison of column bleed from (*a*) a relatively new capillary column with a low column bleed profile from (*b*) a frequently used column of identical dimensions with a higher column bleed profile and film thickness under the same gas chromatographic conditions and from (*c*) a fatigued column used for a separation of a chloroform extract of an air-filtration system at a local social establishment. *(Continued)*

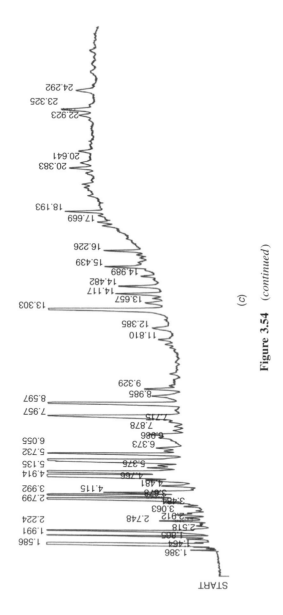

Figure 3.54 (*continued*)

(*c*)

198

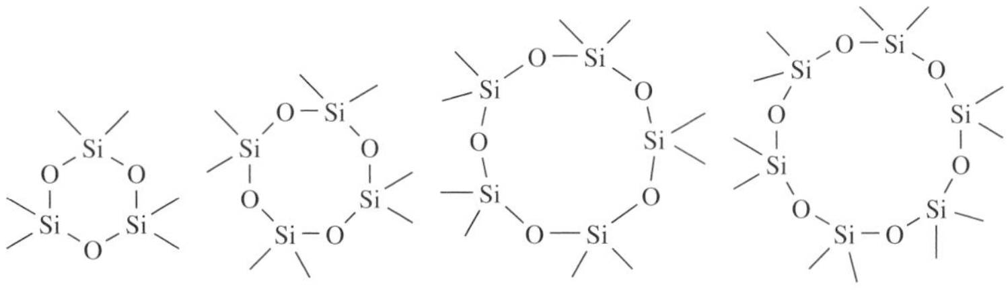

Name:	D3	D4	D5	D6
Molecular mass:	222	296	370	444
Main fragments:	207	281	73, 267, 355	73, 341, 429

Figure 3.55 Structures of cyclic siloxanes produced during the degradation of a polysiloxane.

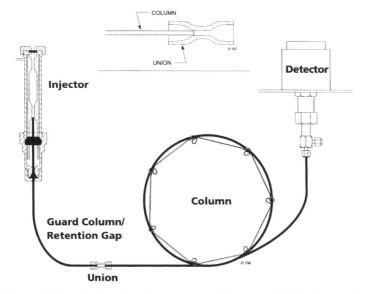

Figure 3.56 Guard column/retention gap. (Courtesy of Walter Jennings.)

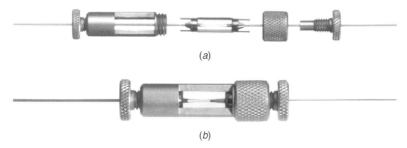

(a)

(b)

Figure 3.57 (a) A disassembled capillary/microbore Vu-Union showing the primary and secondary sealing mechanisms, and (b) an assembled Vu-Union. (Courtesy of the Restek Corporation.)

Vu-Union. The primary and secondary sealing mechanisms in a capillary Vu-Union is shown in Figure 3.57, where the two column ends are positioned into ferrules located inside a deactivated tapered glass insert. This type of connector combines the benefits of a low-dead-volume connection with the sturdiness of a ferrule seal. Furthermore, the glass window permits visual confirmation of the connection.

Recently, guard columns integrated into the overall column as one continuous length have become commercially available; one such configuration is termed an *Integra-Guard Column*. This type of configuration eliminates the need to make a leak-free connection between a guard column and an analytical column and is a convenient alternative in applications and inlet systems requiring a retention gap.

Column Fatigue and Regeneration

Deterioration in column performance can occur by contamination of the column with the accumulation of nonvolatiles and particulate matter, usually in the injector liner and column inlet. Column contamination is manifested by adsorption and peak tailing of active analytes, excessively high column bleed levels, and changes in the retention characteristics of the column. Rejuvenation of the column can be attempted by several paths. First, remove 1 or 2 meters of column from the inlet end. If the column still exhibits poor chromatographic performance, try turning the column around and reconditioning it overnight disconnected from the detector. A third approach is solvent rinsing, an extreme measure that should be attempted only with cross-linked or chemically bonded phases. Solvent-rinse kits such as the one shown schematically in Figure 3.58 are available commercially and enable a column to be back-rinsed of contamination by slowly introducing 10 to 30 mL of an appropriate solvent into the detector end of the column by nitrogen gas pressure. The result of this procedure is shown in Figure 3.59. This approach is worthwhile for heavily contaminated columns, but in all cases the recommendations for rinsing outlined by the column manufacturer should be followed.

3.9   SPECIAL GAS CHROMATOGRAPHIC TECHNIQUES

Simulated Distillation

Simulated distillation is a gas chromatographic procedure employed to determine the boiling range of hydrocarbons in a fuel product; the method commonly employed is ASTM Method D2887. The boiling range up to 538°C specified in the method includes fuels such as diesel, kerosene, gas oil, and jet fuel. Traditionally, this method has been used with packed columns containing a nonpolar stationary phase, but the ASTM method now permits a 0.53-mm-i.d. capillary column to be used. The procedure is fairly straightforward. Calibration is achieved with a mixture of *n*-alkanes of known boiling points to encompass the carbon number range of the sample. Hydrocarbons will elute in order of increasing boiling point on a nonpolar column. With the assistance of computer software, a calibration curve is constructed; this curve is then used for computation of the boiling range of

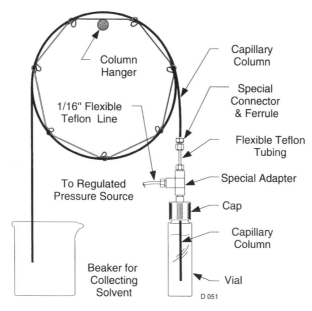

Figure 3.58   Solvent rinse kit. (Courtesy of Walter Jennings.)

the petroleum fraction and the energy Btu) offered by the sample. Basically, the total peak area of a sample is divided into segments or slices, and then the peak area associated with a given segment is converted to a boiling-point range. The computational procedure is analogous to that used for molecular-weight determinations in the size-exclusion mode of HPLC, where the peaks in a chromatogram are digitally sliced into trapezoidal areas, with each area related to a molecular-weight range. This gas chromatographic method, also known as *Sim Dist*, is widely used in the petroleum, automotive, and aircraft industries. Many column suppliers have descriptive materials available on their Web sites on columns used for this methodology.

Multidimensional Gas Chromatography

Despite the highly efficient separations produced with a capillary column, chromatography needs the increased resolution available in the analysis of complex samples via increased peak capacity. Phillips and his students are credited with the construction of the first generation of comprehensive two-dimensional GC by incorporating a transfer device between two columns connected serially (136–139). The term *comprehensive two-dimensional GC*, sometimes abbreviated as C2DGC, is used interchangeably with GC × GC. There is a difference between two-dimensional heartcutting and comprehensive GC × GC, which is illustrated in Figure 3.60. In two-dimensional gas chromatographic separation in the heartcutting mode, two columns are operated in series via a flow-diverting device that allows

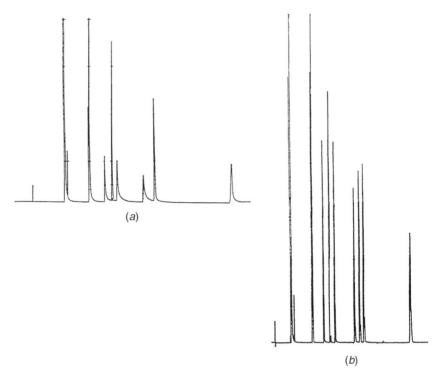

Figure 3.59 Regeneration of a capillary column by solvent rinsing with *n*-pentane: (*a*) fatigued column; (*b*) after rinsing will 30 mL of *n*-pentane.

user-designated segments of the effluent from the first column onto the second column (Figure 3.60*a*). The specific aim of a heartcut procedure is the enhanced resolution needed in regions of a one-dimensional separation.

However, in GC × GC, the effluent from the first column is rapidly sampled and diverted to a second column operated under fast chromatographic conditions. Typically, the second column is more polar and usually shorter in length than the first column. The interface in Figure 3.60*b* is termed a *modulator*, the central item in a GC × GC separation. With each modulation or sampling of the first column effluent, a chromatogram is generated. Most commercial modulators function thermally, requiring large quantities of cryogenic cooling and a gaseous working fluid (140,141). Valve-based modulators have appeared which are simpler in design and more straightforward; there are two main types: subsampling (142,143) and differential flow (144–146). In addition to a modulator, software for the visualization of two-dimensional separations is somewhat sophisticated and is essential. The process is presented sequentially in Figure 3.61, where the raw detector signal is shown in part (*a*); the segmentation or modulation of the sample into the second column is represented in part (*b*) by tick marks on the retention time axis, and a three-dimensional surface plot showing the rearrangement of two-dimensional

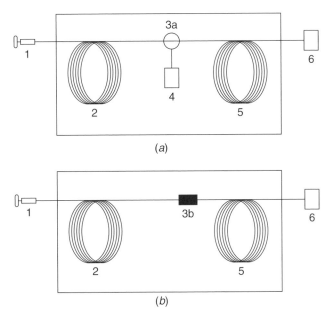

Figure 3.60   Two-dimensional GC systems. (*a*) heartcut GC × GC; (*b*) comprehensive GC × GC. (1) first-dimension inlet; (2) first-dimension column; (3a) diversion valve; (3b) modulator; (4) detector; (5) second-dimension column; (6) second-dimension detector. [From *Anal. Chem.*, 76,167–174A (2004), copyright © 2004 American Chemical Society, with permission.]

frames into a matrix array is shown in part (*c*). With software one can also generate contour plots with spots of specific regions indicated by different colors which enhance the elegance of the technique.

GC × GC-ECD, GC × GC-NPD, and GC × GC-TOFMS have been investigated in numerous studies, and an overview of GC × GC has been published by Dimandja (147). We anticipate substantial growth in applications of this technique.

Computer Modeling of Stationary Phases

Many chromatographers select a column and corresponding dimensions for analysis, then optimize the separation by varying column temperature. Keeping the adage "like dissolves like" in mind, the selection of the stationary phase is often made by (1) experience, (2) examination of Rohrschneider or McReynolds constants, (3) guidance from Web sites of column manufacturers, or (4) use of computer modeling and optimization of selectivity via computer assistance. Studies focusing on predicting separations in GC and selectivity prediction using computer modeling have appeared in the literature (148–150). Dorman and colleagues designed new stationary phases for specific applications using a modeling process with the ultimate separation predicted using thermodynamic relationships (149,150). In an alternative approach, Wick and co-workers studied molecular simulation of gas–liquid

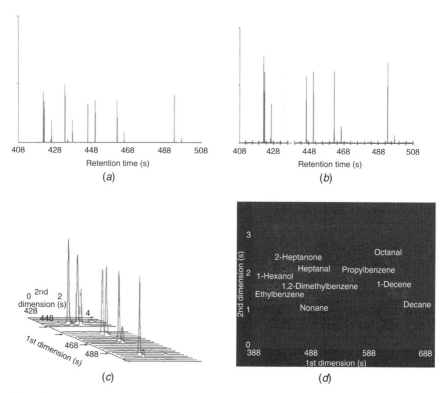

Figure 3.61 GC × GC process. (*a*) Raw detector signal; (*b*) Segmentation into second-dimension chromatograms; (*c*) three-dimensional surface plot showing rearrangement of second-dimension "frames" into a matrix array; (*d*) Contour plot of a larger portion of the GC × GC chromatogram. [From *Anal. Chem.*, 76, 167–174A (2004), copyright © 2004 American Chemical Society, with permission.]

adsorption and partitioning in GLC as well as the effects of solute overloading on retention in GLC (151,152). Commercial software for prediction of chromatographic selectivity and separations is discussed further in Chapter 4.

REFERENCES

1. K. J. Hyver (Ed.) *High Resolution Gas Chromatography*, 3rd ed., Hewlett-Packard, Palo Alto, CA, 1989, Chap. 2.

2. M. J. E. Golay, in *Gas Chromatography 1957* (East Lansing Symposium), V. J. Coates, H. J. Noebels, and I. S. Fagerson (Ed.), Academic Press, New York, 1958, pp. 1–13.

3. M. L. Lee and B. W. Wright, *J. Chromatogr.*, **184**, 234 (1980).

4. R. Dandeneau and E. H. Zerenner, *J. High Resolut. Chromatogr. Chromatogr. Commun.*, **2**, 351 (1979).

5. R. Dandeneau and E. H. Zerenner, *Proc. 3rd International Symposium on Glass Capillary Gas Chromatography,* Hindelang, Germany, 1979, pp. 81–97.

6. M. L. Lee, F. J. Yang, and K. D. Bartle, *Open Tubular Gas Chromatography: Theory and Practice,* Wiley, New York, 1984, p. 4.

7. R. R. Freeman, *High Resolution Gas Chromatography*, 2nd ed., Hewlett-Packard, Palo Alto, CA, 1989, p. 3.

8. D. H. Desty, J. N. Haresnape, and B. H. Whyman, *Anal. Chem.*, **32**, 302 (1960).

9. W. G. Jennings, *Comparisons of Fused Silica and Other Glasses Columns in Gas Chromatography*, Alfred Heuthig, Publishers, Heidelberg, Germany, 1986, p. 12.

10. S. R. Lipsky and W. J. McMurray, *J. Chromatogr.*, **217**, 3 (1981).

11. W. G. Jennings, *Analytical Gas Chromatography*, 2nd ed., Academic Press, San Diego, CA, 1997, p. 36.

12. L. Borksanyl, O. Liardon, and E. Kovats, *Adv. Colloid Interface Sci.*, **6**, 95 (1976).

13. E. R. Lippincott and R. Schroeder, *J. Phys. Chem.*, **23**, 1099 (1955).

14. S. R. Lipsky, W. J. McMurray, M. Hernandez, J. E. Purcell, and K. E. Billeb, *J. Chromatogr. Sci.*, **18**, 1 (1980).

15. J. Macomber, R. Hinz, T. Ewing, and R. Acuna, *LC/GC*, The Application Notebook, June, p. 81 (2005).

16. J. Macomber and P. Nico, *LC/GC*, The Application Notebook, February, p. 55 (2005).

17. J. Macomber, P. Nico, and G. Nelson, *LC/GC*, The Application Notebook, February, p. 66 (2004).

18. J. Macomber, R. Timmerman, and P. Lemke, *LC/GC*, The Application Notebook, June, p. 72 (2004).

19. J. Macomber and L. Begay, *LC/GC*, The Application Notebook, September, p. 72 (2003).

20. J. Macomber, P. Nico, and G. Nelson, *LC/GC*, The Application Notebook, June, p. 63 (2003).

21. J. Macomber and G. Nelson, *LC/GC*, The Application Notebook, June, p. 48 (2002).

22. S. Griffin, *LC/GC*, **20**, 928 (2002).

23. W. Jennings, *Am. Lab.*, **20**, 18 (2002).

24. S. Trestianu and G. Gilioli, *J. High Resolut. Chromatogr. and Chromatogr. Commun.*, **8**, 771 (1985).

25. F. Bruner, M. A. Rezai, and L. Lattanzi, *Chromatographia*, **41**(7–8), 403 (1995).

26. J. Buijten, L. Blomberg, K. Markides, and T. Wannman, *J. Chromatogr.*, **237**, 465 (1982).

27. G. Rutten, J. de Haan, L. van de Ven, A. van de Ven, H. van Cruchten, and J. Rijks, *J. High Resolut. Chromatogr. and Chromatogr. Commun.*, **8**, 664 (1985).

28. C. L. Woolley, K. E. Markides, M. L. Lee, and K. D. Bartle, *J. High Resolut. Chromatogr. Chromatogr. Commun.*, **9**, 506 (1986).

29. C. L. Woolley, R. C. Kong, B. E. Richter, and M. L. Lee, *J. High Resolut. Chromatogr. Chromatogr. Commun.*, **7**, 329 (1984).

30. K. E. Markides, B. J. Tarbet, C. M. Schregenberger, J. S. Bradshaw, M. L. Lee, and K. D. Bartle, *J. High Resolut. Chromatogr. Chromatogr. Commun.*, **8**, 741 (1985).

31. H. Traitler, *J. High Resolut. Chromatogr. Chromatogr. Commun.*, **6**, 60 (1983).

32. W. Blum, *J. High Resolut. Chromatogr. Chromatogr. Commun.*, **9**, 120 (1986).

33. L. J. M. van de Ven, G. Rutten, J. A. Rijks and J. W. de Haan, *J. High Resolut. Chromatogr. Chromatogr. Commun.*, **9**, 741 (1986).

34. M. A. Moseley and E. D. Pellizari, *J. High Resolut. Chromatogr. Chromatogr. Commun.*, **5**, 472 (1982).

35. A. Venema and J. T. Sukkei, *J. High Resolut. Chromatogr. Chromatogr. Commun.*, **8**, 705 (1985).

36. V. Pretorius and D. H. Desty, *J. High Resolut. Chromatogr. Chromatogr. Commun.*, **4**, 38 (1981).

37. B. Xu and N. P. E. Vermulen, *J. High Resolut. Chromatogr. Chromatogr. Commun.*, **8**, 181 (1985).

38. R. C. Kong, C. L. Woolley, S. M. Fields, and M. L. Lee, *Chromatographia*, **18**, 362 (1984).

39. T. Welsch and H. Frank, *J. High Resolut. Chromatogr. Chromatogr. Commun.*, **8**, 709 (1985).

40. K. E. Markides, B. J. Tarbet, C. L. Woolley, C. M. Schrengenberger, J. S. Bradshaw, M. L. Lee, and K. D. Bartle, *J. High Resolut. Chromatogr. Chromatogr. Commun.*, **8**, 378 (1985).

41. K. Grob, *Making and Manipulating Capillary Columns for Gas Chromatography*, Alfred Heuthig Publishers, Heidelberg, Germany, 1986,

42. T. Welsch, *J. High Resolut. Chromatogr.*, **11**, 471 (1988).

43. G. Schomburg, H. Husmann, S. Ruthe, and T. Herraiz, *Chromatographia*, **15**, 599 (1982).

44. C. L. Woolley, K. E. Markides, and M. L. Lee, *J. Chromatogr.*, **367**, 9 (1986).

45. M. Hetem, G. Rutten, B. Vermeer, J. Rijks, L. van de Ven, J. de Haan, and C. Cramers, *J. Chromatogr.*, **477**, 3 (1989).

46. L. Blomberg and K. E. Markides, *J. High Resolut. Chromatogr. Chromatogr. Commun.*, **8**, 632 (1985).

47. T. T. Stark, R. D. Dandeneau, and L. Mering, 1980 Pittsburgh Conference, Atlantic City, NJ, Abstract 002.

48. B. W. Wright, P. A. Peaden, M. L. Lee, and T. Stark, *J. Chromatogr.*, **248**, 17 (1982).

49. C. F. Poole and S. K. Poole, *Chromatography Today*, Elsevier, New York, 1991, p. 147.

50. K. Grob and G. Grob, *J. Chromatogr.*, **125**, 471 (1976).

51. D. A. Cronin, *J. Chromatogr.*, **97**, 263 (1974).

52. L. Blomberg and T. Wannman, *J. Chromatogr.*, **148**, 379 (1978).

53. J. L. Marshall and D. A. Parker, *J. Chromatogr.*, **122**, 425 (1976).

54. K. Grob, G. Grob, and K. Grob, *J. High Resolut. Chromatogr. Chromatogr. Commun.*, **1**, 149 (1978).

55. R. F. Arrendale, R. F. Severson, and O. T. Chortyk, *J. Chromatogr.*, **208**, 209 (1981).

56. H. Yang and W. Chen, International patent CO8L079-04, 2001.

57. D. A. Smith, D.. Salabsky, and M. J. Feeney, paper presented at the Eastern Analytical Symposium, Somerset, NJ, November 2000.

58. U.S. patent application, 09/388,868, 2000.

59. W. Jennings, *Gas Chromatography with Glass Capillary Columns*, 2nd ed., Academic Press, New York, 1980.

60. J. Bouche and M. Verzele, *J. Gas Chromatogr.*, **6**, 501 (1968).

61. K. Grob, G. Grob, and K. Grob, *J. Chromatogr.*, **156**, 1 (1976).

62. E. J. Guthrie and J. J. Harland, *LC/GC*, **12**, 80 (1994).

63. D. W.Grant, *Capillary Gas Chromatography*, Wiley, Chichester, West Sussex, England, 1996, p. 38.

64. K. Grob and G. Grob, *J. High Resolut. Chromatogr. Chromatogr. Commun.*, **2**, 109 (1979).

65. B. W. Wright, P. A. Peaden, and M. L. Lee, *J. High Resolut. Chromatogr. Chromatogr. Commun.*, **5**, 413 (1982).

66. W. Noll, *Chemistry and Technology of Silicones*, Academic Press, New York, 1968, p. 464.

67. J. K. Haken, *J. Chromatogr.*, **300**, 1 (1984).

68. L. Blomberg, *J. High Resolut. Chromatogr. Chromatogr. Commun.*, **5**, 520 (1982).

69. L. Blomberg, *J. High Resolut. Chromatogr. Chromatogr. Commun.*, **7**, 232 (1984).

70. Z. Juvanez, M. M. Schmer, D. F. Johnson, K. E. Markides, J. S. Bradshaw, and M. L. Lee, *J. Microcolumn Sep.*, 3(6), 349 (1991).

71. V. G. Verezkin, Y. A. Levin, and V. P. Mukhina, *J. High Resolut. Chromatogr.*, 17(4), 280 (1994).

72. W. Blum and R. Aicholz, *J. Microcolumn Sep.*, 5(4), 297 (1993).

73. K. Janak, I. Haegglund, L. G. Blomberg, S. G. Claude, and R. Tabacchi, *J. Microcolumn Sep.*, 3(6), 497 (1991).

74. Q. Li and C. F. Poole, *J. Sep. Sci.*, **24**, 129 (2001).

75. M. Verzele and P.Sandra, *J. Chromatogr.*, **158**, 211 (1978).

76. J. A. Hubball, P. R. DiMauro, E. F. Barry, E. A. Lyons, and W. A. George, *J. Chromatogr. Sci.*, **22**, 185 (1984).

77. C. Mandini, E. M. Chambaz, M. Rigaud, J. Durand, and P. Chebroux, *J. Chromatogr.*, **126**, 161 (1976).

78. M. Rigaud, P. Chebroux, J. Durand, J. Maclouf, and C. Mandini, *Tetrahedron Lett.*, **44**, 3935 (1976).

79. K. Grob, *Chromatographia*, **10**, 625 (1977).

80. L. Blomberg, J.Buijten, J. Gawdzik, and T. Wannman, *Chromatographia*, **11**, 521 (1978).

81. L. Blomberg and T. Wannman, *J. Chromatogr.*, **168**, 81 (1979).

82. K. Grob and G. Grob, *J. Chromatogr.*, **213**, 211 (1981).

83. K. Grob, G.Grob, and K.Grob, Jr., *J. Chromatogr.*, **211**, 243 (1981).

84. K. Grob and G. Grob, *J. High Resolut. Chromatogr. Chromatogr. Commun.*, **4**, 491 (1981).

85. K. Grob and G. Grob, *J. High Resolut. Chromatogr. Chromatogr. Commun.*, **5**, 13 (1982).

86. P. Sandra, G. Redant, E. Schacht, and M. Verzele, *J. High Resolut. Chromatogr. Chromatogr. Commun.*, **4**, 411 (1981).

87. L. Blomberg, J. Buijten, K. Markides, and T. Wannman, *J. High Resolut. Chromatogr. Chromatogr. Commun.*, **4**, 578 (1981).

88. P. A. Peaden, B. W. Wright, and M. L. Lee, *Chromatographia*, **15**, 335 (1982).

89. J. Buijten, L. Blomberg, K. Markides, and T. Wannman, *Chromatographia*, **16**, 183 (1982).

90. K. Markides, L. Blomberg, J. Buijten, and T. Wannman, *J. Chromatogr.*, **254**, 53 (1983).

91. B. E. Richter, J. C. Kuei, J. I. Shelton, L. W. Castle, J. S. Bradshaw, and M. L. Lee, *J. Chromatogr.*, **279**, 21 (1983).

92. B. E. Richter, J. C. Kuei, N. J. Park, S. J. Crowley, J. S. Bradshaw, and M. L. Lee, *J. High Resolut. Chromatogr. Chromatogr. Commun.*, **6**, 371 (1983).

93. J. Buijten, L. Blomberg, S. Hoffman, K. Markides, and T. Wannman, *J. Chromatogr.*, **289**, 143 (1984).

94. W. Bertsch, V. Pretorius, M. Pearce, J. C. Thompson, and N. G. Schnautz, *J. High Resolut. Chromatogr. Chromatogr. Commun.*, **5**, 432 (1982).

95. J. A. Hubball, P. DiMauro, E. F. Barry, and G. E. Chabot, *J. High Resolut. Chromatogr. Chromatogr. Commun.*, **6**, 241 (1983).

96. E. F. Barry, G. E. Chabot, P. Ferioli, J. A. Hubball, and E. M. Rand, *J. High Resolut. Chromatogr. Chromatogr. Commun.*, **6**, 300 (1983).

97. E. F. Barry, J. A. Hubball, P. R. DiMauro, and G. E. Chabot, *Am. Lab.*, **15**, 84 (1983).

98. G. Alexander, *J. High Resolut. Chromatogr.*, **13**, 65 (1990).

99. H. Liu, A. Zhang, Y. Jin, and R. Fu, *J. High Resolut. Chromatogr.*, **12**, 537 (1989).

100. R. C. M. de Nijs and J. de Zeeuw, *J. High Resolut. Chromatogr. Chromatogr. Commun.*, **5**, 501 (1982).

101. J. Buijten, L. Blomberg, K. E. Markides, and T. Wannman, *J. Chromatogr.*, **268**, 387 (1983).

102. O. Etler and G. Vigh, *J. High Resolut. Chromatogr. Chromatogr. Commun.*, **8**, 42 (1985).

103. L. Bystricky, *J. High Resolut. Chromatogr. Chromatogr. Commun.*, **9**, 240 (1986).

104. W. A. George, Ph.D. dissertation, University of Massachusetts–Lowell, 1986.

105. J. A. Hubball, Ph.D. dissertation, University of Connecticut, 1987.

106. M. Horka, V. Kahle, K. Janak, and K. Tesarik, *Chromatographia*, **21**, 454 (1985).

107. S. R. Lipsky and W. J. McMurray, *J. Chromatogr.*, **289**, 129 (1984).

108. W. Blum, *J. High Resolut. Chromatogr. Chromatogr. Commun.*, **8**, 718 (1985).

109. W. Blum, *J. High Resolut. Chromatogr. Chromatogr. Commun.*, **9**, 350 (1986).

110. W. Blum, *J. High Resolut. Chromatogr. Chromatogr. Commun.*, **9**, 120 (1986).

111. W. Blum and L. Damasceno, *J. High Resolut. Chromatogr. Chromatogr. Commun.*, **10**, 472 (1987).

112. W. Blum and G. Eglinton, *J. High Resolut. Chromatogr. Chromatogr. Commun.*, **12**, 290 (1989).

113. P. Schmid and M. D. Mueller, *J. High Resolut. Chromatogr. Chromatogr. Commun.*, **10**, 548 (1987).

114. T. Welsch and O. Teichmann, *J. High Resolut. Chromatogr.*, **14**, 153 (1991).

115. F. David, P. Sandra, and G. Diricks, *J. High Resolut. Chromatogr. Chromatogr. Commun.*, **11**, 256 (1988).

116. R. Aichholz, *J. High Resolut. Chromatogr.*, **13**, 71 (1990).

117. K. Grob and G. Grob, *J. Chromatogr.*, **347**, 351 (1985).

118. A. C. Pierre, *Introduction to Sol Processing*, Kluwer, Boston, 1998.

119. D. A. Loy, K. Rahimian, and M. Samara, *Angew. Chem. Int. Ed.*, **38**, 555 (1999).

120. S. A. Rodriguez and L. A. Colon, *Anal. Chim. Acta*, **397**, 207 (1999).

121. Y. Guo and L. A. Colon, *Anal.Chem.*, **67**, 2511 (1995).

122. J. D. Hayes and A. Malik, *J. Chromatogr. B*, **695**, 3 (1997).

123. D.-X. Wang, S.-L. Chong, and A. Malik, *Anal. Chem.*, **69**, 4566 (1997).

124. W. A. Konig, S. Lutz, P. Mischnick-Lubecke, R. Brassat, and G. Wenz, *J. Chromatogr.*, **441**, 471 (1988).

125. D. W. Armstrong and W. Y. Li, *Anal. Chem.*, **62**, 217 (1990).

126. V. Schurig and H. Nowotny, *J. Chromatogr.*, **441**, 155 (1988).

127. V. Schurig, D, Schmalzing, U. Muhleck, M. Jung, M. Schleimer, P. Mussche, C. Duvekot, and J. C. Buyten, *J. High Resolut. Chromatogr.*, **13**, 713 (1990).

128. J. V. Hinshaw, *LC/GC*, **11**, 644 (1993).

129. *A Guide to Using Cyclodextrin Bonded Phases for Chiral Separations by Capillary Gas Chromatography,* Advanced Separations Technologies, Whippany, NJ, 2002.

130. J. Ding, T. Welton, and D. W. Armstrong, *Anal. Chem.*, **76**, 6819 (2004).

131. R. C. M. de Nijs, *J. High Resolut. Chromatogr. Chromatogr. Commun.*, **4**, 612 (1981).

132. J. de Zeeuw, R. C. M. de Nijs, J. C. Buijten, J. A. Peene, and M. Mohnke, *Am. Lab.*, **19**, 84 (1987).

133. Z. Ji, R. E. Majors, and E. J. Guthrie, *J. Chromatogr. A*, **842**, 115 (1999).

134. J. de Zeeuw and M. Barnes, *Am. Lab.*, November 2005.

135. R. Bartram, in *Modern Practice of Gas Chromatography,* 4th ed., Wiley, Hoboken, NJ, 2004, Chap. 10.

136. J. B. Phillips, D. Luu, and R. Lee, *J. Chromatogr. Sci.*, **24**, 386 (1986).

137. S. Mitra and J. B. Phillips, *J. Chromatogr. Sci.*, **26**, 620 (1988).

138. Z. Liu and J. B. Phillips, *J. Microcolumn Sep.*, **1**, 249 (1989).

139. Z. Liu and J. B. Phillips, *J. Chromatogr. Sci.,* **29**, 227 (1991).

140. E. B. Ledford and C. Billesbach, *J. High Resolut. Chromatogr.*, **23**, 202 (2000).

141. J. Beens, M. Adahchour, R. J. J. Vreuls, K. van Altenaand, and U. A. T. Brinkman, *J. Chromatogr. A*, **919**, 127 (2001).

142. C. A. Bruckner, B. J. Prazen, and R. E. Synovec, *Anal. Chem.*, **70**, 2796 (1998).

143. J. F. Hamilton, A. C. Lewis, and K. D. Bartle, *J. Sep. Sci.*, **26**, 578 (2003).

144. J. V. Seeley, F. Kramp, and C. J. Hicks, *Anal. Chem.*, **72**, 4346 (2000).

145. P. A. Bueno and J. V. Seeley, *J. Chromatogr. A*, **1027**, 3 (2004).

146. J. W. Diehl and F. P. Di Sanzo, *J. Chromatogr. A*, **1080**, 157 (2005).

147. J.-M. D. Dimandja, *Anal. Chem.*, **10**, 167A (2004).

148. F. L. Dorman, P. D. Schettler, C. M. English, and D. V. Patwardhan, *Anal. Chem.*, **74**, 2133 (2002).

149. F. L. Dorman, C. M. English, M. J. Feeney, and J. Kowalski, *Am. Lab.*, **31**, 20 (1999).

150. F. L. Dorman, P. D. Schettler, C. M. English, and M. J. Feeney, *LC/GC*, **18**, 928 (2000).

151. C. D. Wick, J. I. Siepmann, and M. R. Schure, *Anal. Chem.*, **74**, 37 (2002).

152. C. D. Wick, J. I. Siepmann, and M. R. Schure, *Anal. Chem.*, **74**, 3518 (2002).

4 Column Oven Temperature Control

4.1 THERMAL PERFORMANCE VARIABLES AND ELECTRONIC CONSIDERATIONS

Gas chromatographic columns are installed in a column oven in which the temperature must be controlled accurately and precisely, because column temperature has a pronounced influence on retention time. Any fluctuation in column temperature will yield an impact on the measurement of retention data and retention indices. Present oven geometries and electronic temperature control components are capable of thermostatting a column oven to $\pm 0.1°C$.

There are several additional requirements that a column oven must satisfy. A column oven should be thermally insulated from heated injector and detector components, a requirement that becomes more demanding as the column oven temperature selected approaches ambient temperature. Ideally, the temperature of a column oven should remain constant and independent of environmental changes in the laboratory and any line-voltage fluctuations. Versatility in the operating temperature capability is also necessary to achieve column temperatures ranging from subambient temperature to elevated temperatures above $400°C$ (for separations with metal-clad capillary columns). With recent advances in adsorbents and PLOT columns, the need for cryogenic cooling of a column oven for subambient separations of permanent gases and light hydrocarbons is no longer required but has been replaced by the need of cryogenic capability for solute focusing purposes with on-column injection and auxiliary sampling techniques, such as thermal desorption, purge and trap, and large volume injection.

Current gas chromatographic oven design is also a product of the age of miniaturization. Early column ovens were relatively large in volume (up to 3500 in³) to accommodate U-shaped glass columns for biomedical and preparative separations. Thermal gradients were common in these huge vertically configured rectangular ovens. On the other hand, the typically smaller oven geometry of today (i.e., $30 \times 27 \times 15$ cm or approximately 12 L) can comfortably accommodate two capillary columns, a packed and a capillary column, or two packed columns. Forced-air convection is the most popular type of gas chromatographic oven, because it provides a uniform temperature in the column oven. Modern oven designs also

Columns for Gas Chromatography: Performance and Selection, by Eugene F. Barry and Robert L. Grob
Copyright © 2007 John Wiley & Sons, Inc.

permit fast cool-down rates after temperature programming, an important consideration because it governs sample throughput in a laboratory.

In modern gas chromatographs the temperature controller of a column oven is a microprocessor incorporated into a feedback loop, allowing both temperature-programming ramp profiles and isothermal heating to be accomplished accurately and reproducibly. Under microprocessor control, a flap or door movement permits blending the proper amount of ambient lab air with oven air in the control of oven temperature. In addition, a cryogenic value can be opened by a microprocessor for delivery of carbon dioxide or liquid nitrogen in the column oven.

An alternative column-oven technology has recently been introduced by Antek Instruments, which manufactures a microwave oven for GC (www.antekhou.com/product/chrom/oven.htm). By heating only the gas chromatographic column rather than the entire column oven, the module provides rapid heating rates ($10°C/s$) and superior cooling rates (greater than $300°C/s$), thus minimizing heating and cool-down times and therefore decreasing the time of analysis and increasing the sample throughput in a laboratory. Temperature-programming rates ranging from $1°C/min$ to $600°C/min$ renders this oven capable of achieving "fast" chromatographic separations, discussed in Section 4.5. This mode of column heating easily converts a conventional one-analysis-at-a-time gas chromatograph into a dual analyzer that can use concurrent yet different temperature programs, doubling the gas chromatographic capabilities of a laboratory. This microwave oven may also be purchased as a retrofit to an existing gas chromatograph.

4.2 ADVANTAGES OF TEMPERATURE PROGRAMMING OVER ISOTHERMAL OPERATION

Isothermal operation of a chromatographic column has a number of drawbacks, as illustrated in the scenario depicted for the separation of lime oil in Figure 4.1. If the isothermal column temperature selected is too low, the early eluting peaks will be closely spaced, while the more strongly retained components will be broad and low-lying (Figure 4.1A). These strongly retained components can be eluted more quickly by selecting a higher isothermal temperature, which will also improve their detectability (Figure 4.1B). However, in doing so, more rapid coelution of components, peaks too closely spaced, and an overall loss in resolution result in the beginning of a chromatogram. This situation, which prevails in all practiced versions of elution chromatography, is often called the *general elution problem*; it is solved in GC by temperature programming, where the column oven temperature is gradually increased at a linear rate during an analysis (Figure 4.1C).

Temperature programming offers several attractive features. One can expect a reduced time of analysis and improved overall detectability of components (peaks are sharper and have nearly equal bandwidths throughout the chromatogram). In the case of unknown samples or samples of high complexity, high-boiling components, which might not elute or be detected under isothermal conditions, can exhibit a more favorable retention time. Temperature programming also helps "clean out" a column

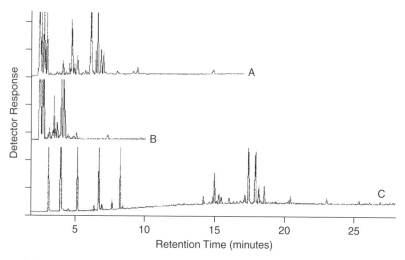

Figure 4.1   Gas chromatographic version of the general elution problem with a separation of lime oil. Curve A, isothermal, 150°C; B, isothermal, 180°C; C, 50 to 250°C at 6°C/min.

of remnant high-boiling species from previous injections. The interested reader is urged to consult the classic book *Programmed Temperature Gas Chromatography* by Harris and Habgood (1) for a detailed treatment of the subject.

4.3   OVEN TEMPERATURE PROFILES FOR PROGRAMMED-TEMPERATURE GAS CHROMATOGRAPHY

Three basic types of temperature programming profiles are used in GC: ballistic, linear, and multilinear. *Ballistic programming* occurs when an oven maintained at a given isothermal temperature is changed rapidly to a much higher isothermal temperature (Figure 4.2*a*) and is sometimes used for fast conditioning of a gas–solid chromatographic column after it has been unused for a period of time. More commonly, programming of this type is incorporated into peripheral sampling methods to quickly drive off solutes from an adsorbent. An example is the purge-and-trap procedure for the determination of volatiles in aqueous solution, where collected solutes are thermally desorbed by ballistic programming (and also by rapidly controlled linear temperature programming) from a silica gel/charcoal/Tenax trap so that they migrate as a narrow zone to the inlet of a capillary column where they are focused. A ballistic ramp may also be used with cryogenic solute focusing to elevate the column temperature quickly to above ambient temperature. However, a chromatographic column maintained at an elevated temperature, then ballistically programmed, can suffer severe damage due to disruption of the stationary-phase film caused by this thermal shock.

The most widely used temperature program is the *linear profile*, shown in Figure 4.2*b*. Here the run begins at a low initial temperature, which may be

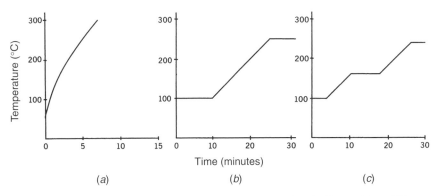

Figure 4.2   Types of temperature programming: (*a*) ballistic; (*b*) linear; (*c*) multilinear.

maintained for a certain number of minutes (an isothermal hold), after which the column oven temperature is raised at a linear rate to the selected final temperature, where it can also be maintained for a specific time interval. The initial temperature and hold period are usually determined from a scouting run made while noting elution temperatures; proper selection of the initial conditions will permit the separation of the low boilers in the separation, while the final temperature chosen should be sufficient for the elution of the more strongly retained components in the sample (keeping in mind the upper temperature limit of the stationary phase). *Multilinear profiles* (Figure 4.2*c*) may be employed in some instances to fine-tune or enhance the resolution in a separation, but are used more commonly

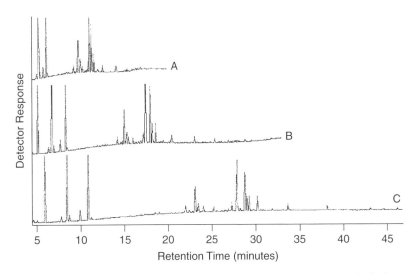

Figure 4.3   Effect of rate of temperature programming on resolution and analysis time; column 30 m × 0.25 mm i.d. DB-1 (0.25-μm film); split injection(100:1), 27 cm/s He, detector FID. Curve A, 12°C/min; B, 6°C/min; C, 3°C/min.

in conjunction with on-column injection. In this injection mode, a low column temperature is maintained during sample introduction into the retention gap, then initiation of the first and usually faster ramp rate induces the solvent and components to start moving in the analytical column, and the final ramp is implemented for elution of the components.

The ramp rate governs the trade-off between analysis time and resolution. The compromise between resolution and time of analysis is contrasted in Figure 4.3 for parallel capillary column separations of lemon oil generated at three different programming rates. Relying on experience and intuition in establishing optimum column temperature conditions can be time consuming and inefficient. An alternative route is the use of computer simulation for method development; this approach is discussed in greater detail in Section 4.4.

4.4 ROLE OF COMPUTER ASSISTANCE IN OPTIMIZING SEPARATIONS IN GAS CHROMATOGRAPHY

Optimization procedures usually focus on establishing either the minimal time necessary for acceptable resolution or maximum resolution within a given period of time. A thorough understanding of which parameters *and* how the parameters affect the chromatographic analysis is helpful in optimizing a separation. Typically, resolution and time required are functions of several interrelated column parameters, such as internal diameter, length, stationary phase, and film thickness, as well as operational parameters (i.e., carrier gas velocity and column temperature), as discussed in Chapter 3. Column parameters are normally considered at the beginning when choosing a column for analysis. Operational parameters, on the other hand, are changed during the optimizing process and are typically the primary targets for optimization, particularly column temperature conditions. Hinshaw has detailed the optimization of separations for GC (2).

The use of software for rendering computer assistance in optimizing separations is attractive for several reasons. First, computer assistance facilitates greatly method development more efficiently and rapidly, a very important consideration for analyses requiring temperature programming. It permits more efficient use of the time of the chromatographer and use of gas chromatographic instrumentation. Moreover, one can gather more knowledge about the separation as well as prediction of the separation under different thermal conditions.

There are software packages available for computer assistance in method development, and several of these are discussed here. Each software package is elegant in terms of its capability. Unfortunately, space does not permit full coverage of the strength of each piece of software; nevertheless, we highlight next some features of each.

DryLab (LC Resources)

DryLab software is just one of the computer simulation software packages available for optimization and method development for GC. The initial version of DryLab focused on the optimization of separations in HPLC, where effects of changes in mobile-phase composition in gradient elution HPLC on solute retention were modeled. These efforts were later extended to GC, where the effects of column temperature during temperature programming of solute retention in GC are modeled in much the same manner. Computer simulation allows the prediction of isothermal and temperature-programmed GC as a function of experimental conditions.

In the GC mode, two initial temperature-programmed runs conducted over the same temperature range (i.e., the same initial and final temperatures) at two different heating rates are required for initial input data. For the best combination of predictive range and accuracy, these rates should differ from one another by a factor of about 3. A measured value for t_m and an average value of the column plate number must be entered as well. DryLab software then uses these input data to calibrate the model and consequently, simulating separations of other conditions.

The starting point for method development in DryLab is display of the resolution map. This map illustrates the resolution of a given critical band pair over a range of heating rates of linear temperature programming. This map establishes the feasibility of the separation, identifies the optimum heating rate, and provides warning of potential problems with robustness. Feasibility is observed from critical resolution values shown on the y-axis of the resolution map. Critical resolution value of less than 1.5 indicates low probability of successful analysis since complete separation requires a resolution value of 1.5. The highest point on the resolution map between the extrapolation confidence limits identifies the optimum heating rate of prosperous analysis. The steep fall-off in critical resolution to either side of the optimum indicates a separation that is susceptible to errors resulting from heating rate. The optimum of heating is identified with the best combination of critical resolution and speed of analysis. Complete conditions, including initial temperature, final temperature, and heating rate, can be modified and evaluated in the gradient editor of DryLab software. Separations by multilinear temperature programming and isothermal conditions (where heating rate is set to zero) can be simulated as well. Separation conditions either isothermally for a simple mixture or temperature programming for a complex mixture can be evaluated easily; applications of this software have been published (3–7); further details can be obtained at www.lcresources.com.

ProezGC (Restek Corporation)

The fundamental concept of ProezGC involves thermodynamic retention indices, which are constants for specific compounds analyzed on specific stationary phases. Since the thermodynamic retention indices are independent of column configuration or analysis conditions, these can be varied by the software to give very accurate estimates of retention times. Therefore, even if the column length, i.d., and film thickness, or flow rate and oven temperatures, are varied, the retention times can be modeled with very high accuracy ($<2\%$ versus actual). The only variable that cannot be adjusted is the polarity of the phase. The procedure for optimization involves a sequence of selections.

Once the thermodynamic retention indices have been determined, the software can be used to predict the actual retention times under many sets of analysis conditions. Thermodynamic retention indices are generated by running specific compounds on a specific phase using both a slow and a fast temperature program. These data can be entered into ProezGC to calculate the thermodynamic retention indices, although Restek has generated a significant number of thermodynamic retention indices for a wide variety of compounds on many different phases. Known as *Restek libraries*, these are included with the software to allow the user to model analyses without having to run compounds under fast and slow conditions. If the compounds are not in the library, the customer can generate a user library of their own compounds and allow the software to calculate the thermodynamic retention indices for these compounds.

In modeling a separation one may change the type of carrier gas, and select whether a single column, a dual column, or coupled columns will be modeled. Libraries are listed by type of phase and type of compound (Rtx-1 Volatiles, Rtx-5 Pesticides, Stabilwax FAMEs, etc). After making these selections, one can then select column dimensions (length, i.d., film thickness) and outlet pressure, the latter an important issue in GC-MS, where the components migrate thorough the column not only by carrier gas flow but also pulled by vacuum of the mass spectrometer.

At this point one may select type of flow: constant pressure, constant linear velocity, dead time, inlet pressure, and flow rate. One may also elect to model a range of flow-related parameters, including pressure programming. Column temperature is then selected, which may be either a specific set of temperature-programming conditions or a range of starting temperatures, initial hold times, temperature ramps, final temperatures, and final hold times. Multiple temperature ramps can also be explored. After values of resolution and peak widths are decided, modeling will begin, after which the top 50 solutions will be displayed, ranked first by the number of components resolved, then by total analysis time. Further details can be found in the vendor's catalog or at www.restek.com.

GC-SOS (Chem SW)

GC-SOS (Gas Chromatography Simulation and Optimization Software) is a software application for the simulation and optimization of capillary gas chromatographic separations. Designed to be flexible, it is intended to provide maximum utility to the user with minimal input. With GC-SOS one can develop high-efficiency capillary gas chromatographic methods on a PC. Off-line simulations can be accurately performed in seconds to determine the effects of varying temperature, pressure, and/or column size conditions on separations. The time and effort required to fully optimize separation conditions and method development is reduced dramatically from hours to minutes.

To simulate a separation the user simply enters the run conditions and retention data from as little as one actual data input run. Chromatogram(s) can be imported from a number of popular formats directly into GC-SOS, and a specialized auto-optimization feature can be used to find an optimized set of run conditions for your separation. Temperature programming, pressure programming, and/or column size effects are modeled. Flexible input requirement standard technology (FIRST) enables simulations to be made from as few as one or up to three data input runs.

Flow calculations can be performed, and an animation feature shows the chromatographic separation process. Automatic calculation of several parameters associated with a separation can be generated. In addition, GC-SOS will calculate the effects of temperature-programming conditions, pressure-programming conditions, and changes in column dimensions. Accurate predictions of peak retention times and peak widths are made almost instantly, including automatic calculations of interpeak resolution and plate numbers to assist in the analysis of separations. Flow calculations are included to assist in the location of optimal pressure values. The auto-optimization routine is included to assist by automatically determining an optimized set of temperature and pressure conditions that will give the desired resolution between all peaks in the shortest amount of time.

One to three actual runs must be performed using the gas chromatographic separation that is to be modeled. The ability to use one to three runs is provided to give flexibility to the user. All data input runs must be carried out isobaric conditions (with constant column head pressure). There are no temperature restrictions, and the runs may be isothermal, or consist of up to two ramp profiles and three isothermal segments.

If two data input runs are used, the only requirement is that they differ significantly in temperature conditions. In general, the calculation error for a particular compound decreases as the retention time difference between the two runs increases. For example, if data input run 1 was run isothermally at 100°C, and data input run 2 was run isothermally at 105°C, the retention differences between the two runs would be rather small and may introduce more error that if data input run 2 was run isothermally at 150°C. The two runs can be any combination of isothermal and/or ramp segments as needed, and there is no requirement that either run be made under the same temperature range or the same number of isothermal and ramp segment combinations.

A third data input run can be entered, if desired, for more accurate pressure calculations. It needs to be performed under the same temperature conditions as one of the other two runs, except at a different column head pressure. The information GC-SOS requires from the data input runs is at the minimum-running conditions, retention times for each compound, and column holdup time. If desired, peak width and area inputs may be entered; if not, GC-SOS supplies a default value for these. GC-SOS requires that column dimension data be supplied for all flow calculations. Additional descriptive information can also be entered for convenient record-keeping purposes and may also be found at www.chemsw.com.

4.5   FAST OR HIGH-SPEED GAS CHROMATOGRAPHY

Areas of intense research interest include the entire spectrum of fast or high-speed gas chromatography (HSGC) (8), tunable column selectivity for HSGC and GC/MS

(9), and the use of tandem columns with pressure-tunable selectivity (10,11). Sandra et al. (12) have reviewed the role of column selectivity in GC and the means for controlling selectivity. Publications considering high-speed temperature programming (13,14) and inlet systems (15–17) should be consulted. A special issue of *High Resolution Chromatography*, which featured comprehensive 2DGC, contains considerable information on instrumentation and issues related to HSGC (18).

Time of analysis in chromatography is defined by the retention time t_{Rl} for the last target-component peak to elute from the column:

$$t_{Rl} = \frac{u}{L}(k_l + 1) \tag{4.1}$$

where u is the average carrier gas velocity, L the column length, and k_l the retention factor for the last compound. If the column length used for an analysis is reduced by a factor of 4 and the carrier gas velocity increased by a factor of 6, the analysis time is reduced by a factor of 24. For example, a 6-m-long column operated with hydrogen at an average velocity of 200 cm/s has a holdup time of 3.0 s, and the analysis time for the isothermal analysis of a mixture with a retention factor range of 0 to 10 is 33 s.

As pointed out by Sacks (19), there are two major difficulties for the truly successful use of HSGC:

1. Conventional instruments for GC are inadequate because of extracolumn sources of band broadening, injection plugs that are too wide in space and in time, and other factors. Gross and colleagues (20) described a synchronized dual-valve injection for high-speed GC and suggest that if properly implemented, fast temperature programming coupled with this type of inlet may lead to very large peak capacities n_p (number of perfectly spaced peaks that will fit in a chromatogram with a specified resolution R_s) for approximately 1-s separations

2. Peak capacity is reduced with shorter columns, as described by

$$n_p = 1 + \frac{\sqrt{L/H}}{4R_s} \ln \frac{t_{Rl}}{t_M} \tag{4.2}$$

where t_M is the dead time for the column and H is the height equivalent to a theoretical plate for the column, defined in Chapter 2. Thus, if separation time is reduced by a factor of 4 by a corresponding reduction in column length, the peak capacity is reduced by a factor of 2. This loss in peak capacity makes it more problematic to apply fast gas chromatographic methods to very complex mixtures.

Rapid analysis of mixtures spanning a wide boiling point range can be accomplished only with high-speed temperature programming. Only recently has equipment become available commercially for high-speed temperature programming, up to 1000°C/min. With these conditions, a program from 50°C to 380°C (e.g., near the upper temperature limit of a cross-linked 100%-polydimethylsiloxane fused-silica column) is completed in 20 s. Although convection ovens for conventional GC have changed relatively little over several decades, better temperature controllers and higher-power heaters have resulted in modest increases in maximum linear heating rates. However, over a wide temperature range (50 to 350°C), 50°C/min, is about the highest ramp rate that is linear over the entire temperature range, but higher linear ramp rates, up to about 100°C/min, can be obtained at lower temperatures.

Most conventional applications in GC use temperature-programming rates that are smaller than need be, resulting in substantially longer analysis times with little increase in column resolving power relative to that obtained with faster temperature programming and shorter analysis times. This is a result of the unavailability of necessary instrumentation optimized in component design with acceptable simplicity and performance levels. Sacks has prepared a definitive treatment of fast GC, including all instrumental aspects (19).

Oven cooling is a major problem for fast GC with convection-oven instruments. Convection ovens typically have large thermal mass, and cooling times are several minutes or more. This poses major limitations on instrument cycle time. Some instruments have provisions for more rapid cryogenic cooling with carbon dioxide or liquid nitrogen. However, even in these cases, oven cooling may dominate instrument cycle time for HSGC. Ovens of low thermal mass can greatly improve cycle time and thus sample throughput in a laboratory.

Selectivity Tuning

Capillary columns with different stationary phases can be operated in series to obtain tunable selectivity. The length ratio of two columns, for example, can be

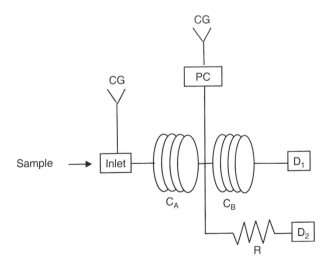

Figure 4.4 Pressure-tunable column ensemble. Columns C_A and C_B use different stationary phases. Electronic pressure controller PC is used to adjust the carrier gas pressure at the column junction point. Detector D_1 monitors the final chromatogram from the column ensemble, and D_2 monitors a small fraction of the effluent from C_A. Carrier gas is provided at points CG. (From ref. 19. Reprinted with permisson of John Wiley & Sons, Inc.)

altered to change the contributions that the two columns make to the overall selectivity (k_o values) in the same way that the volume ratios of the stationary phases
can be changed with mixed-phase packed columns, the window diagram approach
reported by Laub et al. (21–23). A more convenient way to adjust the selectivity
of a series-coupled (tandem) column combination is to provide adjustable carrier gas pressure at the column junction point (10,11,24–26). This is illustrated in
Figure 4.4. An electronic pressure controller PC is used to control the junction-
point pressure, while detector D_1 monitors the output from the tandem column
ensemble. Detector D_2 monitors a small fraction of the effluent from C_A, whereas
capillary restrictor R controls the amount of the C_A effluent directed to D_2. This
second detector is optional, but it is useful for method development.

Changing the carrier gas pressure at the column junction point changes the
residence times of all components in the two columns. For example, if the junction-
point pressure is increased, the pressure drop along C_A decreases and the drop along
C_B increases. This results in decreased carrier gas flow and thus longer solute residence times in C_A. Carrier gas flow increases in C_B, resulting in shorter residence
times. This increases the contribution that C_A makes to the overall selectivity of
the column combination. Figure 4.5 shows a portion of high-speed chromatograms

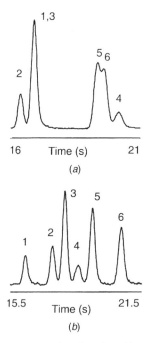

Figure 4.5 Portions of chromatograms showing the effects of a change in the junction-
point pressure using the apparatus of Figure 4.4. The column ensemble consists of a 5.0-m-
long 0.25-mm-i.d. dimethylpolysiloxane column (*a*) followed by a 5.0-m-long 0.25-mm-i.d.
polyethylene glycol column (*b*). (From ref. 19. Reprinted with permisson of John Wiley &
Sons, Inc.)

containing peaks of six components, eluting within 6 s. The only difference between the two chromatograms is the carrier gas pressure at the column junction point. Note that the pattern of peaks is very different for the two junction-point pressures, and for chromatogram *(b)*, complete separation is achieved.

Improved separation quality for more complex mixtures often can be obtained with programmable selectivity where the junction-point pressure of the tandem column ensemble is changed during a separation (27). Initially, the junction-point pressure is set to give optimal separation conditions for the first group of peaks to elute from the column ensemble. After these component bands have migrated into the second column, the junction-point pressure can be changed to a value more appropriate for the next group of components, which are still in the first column. This process can be repeated as many times as necessary. The limitation of tunable and programmable selectivity with tandem capillary columns using electronic pressure control at the column junction point is that changing the junction-point pressure results in a change in the ensemble elution pattern for the entire mixture or a subgroup of the mixture, and under optimal conditions, any pressure change used to enhance the resolution of one component pair will degrade the resolution of another component pair.

A solution to this problem is to replace the electronic pressure controller with a low-dead-volume computer-controlled valve and a source of carrier gas at some preset pressure. Normally, the valve is closed, and the column junction-point pressure is the natural pressure that occurs at the column junction point in the absence of any additional connections. When the valve is opened, the junction-point pressure assumes the preset value of the additional carrier gas source. Usually, the valve is open for only a few seconds, to enhance the resolution of a particular component pair. Thus, the carrier gas in the two columns undergoes a pulsed-flow modulation (28,29). A particularly attractive version of pulsed-flow modulation uses the gas chromatographic inlet pressure as the preset pressure (30,31). Thus, when the valve is opened, both ends of column C_A are at the same pressure, and carrier gas flow in C_A stops (stop-flow operation). Stop-flow operation is used to enhance the resolution of a targeted component pair without significantly changing the elution pattern and resolution of other components in the mixture. The concept is illustrated by the band trajectory plots shown in Figure 4.6 for a pair of components labeled 1 and 2 that are completely separated by the first column but coelute from the column ensemble. The solid-line plots are for the case without a stop-flow pulse, and the dashed-line plots for the case with a 5-s-wide pulse occurring at the time indicated by the vertical lines.

For case *(a)*, the pressure pulse is applied when both components are in C_A. Both bands stop during the pulse, and the peaks in the ensemble chromatogram are shifted to a later time but without significant change in resolution. For case *(b)*, the pulse is applied after the first of the component bands has migrated across the column junction point and is in C_B, but the band for the other component is still in the first column. The band in C_A stops for the duration of the pulse, while the band in C_B migrates more rapidly during the pulse. The result is complete separation of the components in the ensemble chromatogram. For case *(c)*, the pulse is applied

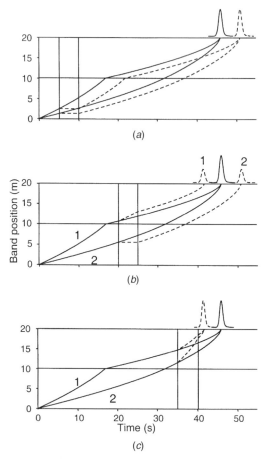

Figure 4.6 Plots of band position versus time for a two-component mixture (1 and 2) illustrating stop-flow operation. Solid-line plots are for a case without a stop-flow pulse, and dashed-line plots are for a case with a 5-s-wide stop-flow pulse, indicated by the vertical lines. (*a*) Stop-flow pulse applied when both components are in C_A; (*b*) stop-flow pulse applied when component 1 is in C_B and component 2 is in C_A; (*c*) stop-flow pulse applied when both components are in C_B. (From ref. 19. Reprinted with permisson of John Wiley & Sons, Inc.)

after both components have crossed the junction and are in C_B, and both peaks are shifted to shorter retention times but with no significant change in resolution.

Figure 4.7 shows the high-speed separation of a 20-component pesticide mixture (plus one impurity peak). The 14-m-long 0.18-mm-i.d. thin-film column ensemble consists of 7.0 m of a trifluoropropylmethyl polysiloxane column followed by 7.0 m of 5% phenyldimethylpolysiloxane column segment. For chromatogram (*a*), no stop-flow pulses were used, and component pairs 2,3 and 10,11 coelute. For chromatogram (*b*), a single 2-s-wide stop-flow pulse was used to enhance the

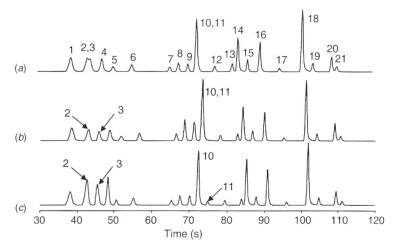

Figure 4.7 High-speed temperature-programmed separation of a 20-component pesticide mixture without stop-flow operation (*a*), with a 2-s-wide stop-flow pulse to separate components 2 and 3 (*b*), and with two stop-flow pulses to separate component pairs 2,3 and 10,11 (*c*). Components are (1) α-BHC; (2) β-BHC; (3) γ-BHC, (4) δ-BHC; (5) heptachlor; (6) aldrin; (7) heptachlor eopxide; (8) α-chlordane; (9) γ-chlordane; (10) 4,4'-DDE; (12) dieldrin; (13) endrin; (14) 4,4'-DDD; (15) endosulfan II, (16) 4,4'-DDT; (17) endrin aldehyde; (18) metoxychlor; (19) endosulfan sulfate; (20) impurity; (21) endrin ketone. (From ref. 19. Reprinted with permisson of John Wiley & Sons, Inc.)

resolution of peak pair 2,3. Note that the peak pattern and resolution of the other components shows no significant change. For chromatogram *(c)*, a second stop-flow pulse was added to enhance resolution of component pair 10,11. With the two stop-flow pulses, complete separation is achieved in about 110 s. A stop-flow system for EPC with either split/splitless or cool on–column inlet modes is commercially available (Restek). Wittrig et al. (32) have described the simple use and merits of the technique.

Fast GC X GC with short primary columns has also been achieved (33). The interested reader is urged to consult reference 19 as well as the publications cited in this section for a further understanding of the separation capabilities of fast GC.

Resistively Heated Columns and Column Jackets

These topics encompass separations achieved with a conventional gas chromatograph without appreciable inlet and/or pneumatic modifications, as outlined earlier. Here we discuss how fast gas chromatographic separations can be achieved with a conventional gas chromatograph using one of the following:

1. Separations achieved with a dedicated gas chromatograph for fast GC (flash gas chromatograph) with a much shorter length of capillary column inserted into a resistively heated column jacket

2. A supplemental resistively heating element residing on the floor of the column oven contributing to the overall column heating (the GC Racer)

3. Chromatographic column installed in a module resistively heated and mounted on the door of a conventional gas chromatograph (LTM module)

We shall now discuss the various paths for obtaining fast gas chromatographic separations. In Figure 4.8a a flash gas chromatograph manufactured by Thermedics

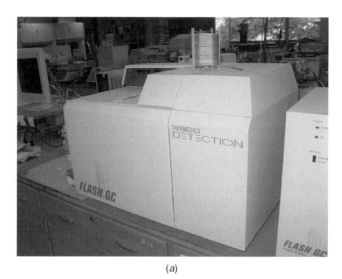

(a)

(b)

Figure 4.8 (a) Flash gas chromatograph (note the gas chromatographic column on the top of the instrument; (b) resistively heated column. Rapid heating is achieved with electrical heating of the column jacket.

Detection is shown as well as a typical column compatible with this chromatograph (Figure 4.8*b*). One should note the metallic terminations at the column end in Figure 4.8*b* serving as the point of contact for rapid, electrical resistive heating. Replacement of a column is straightforward; the old column is pulled out and the new column is inserted gently through the jacket. A comparison of separations of

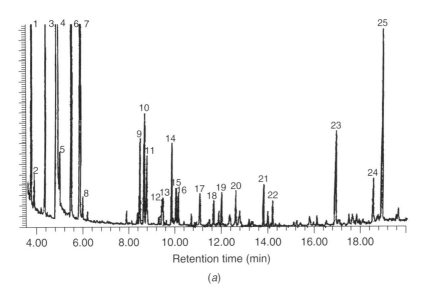

(a)

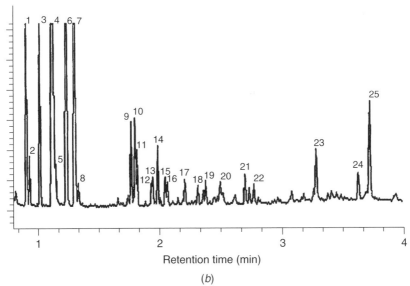

(b)

Figure 4.9   Separations of tangerine oil: (*a*) conventional GC: temperature programming from 50 to 240°C at 9°C/min; (*b*) high-speed GC: 50 to 240°C at 60°C/ min.

tangerine oil under conventional and high-speed temperature-programming conditions is illustrated in Figure 4.9, where reduction in analysis time greater than a fourfold may be noted.

Resistively heated column effects can also be achieved by installing a secondary or auxiliary heating supply in the chromatographic oven (Figure 4.10); the device is

Figure 4.10   Oven-heating module of GC Racer on the floor of a column oven compartment.

Figure 4.11   GC Racer controller.

commercially available from Zip Scientific or Restek (www.restekcorp.com) and is named the GC Racer (Figure 4.11) The heating element of the secondary supply is connected to the controller plugged into the main PC board of the chromatograph. When fast temperature-programming rates are desired, the heating element of the Racer is fed into the circuit electrically to augment the heat being supplied to the column oven. The GC Racer system will maintain temperature program rates of 50°C/min up to 350°C and 60°C/min to temperatures as high as 450°C. Parallel chromatograms of a diesel fuel generated at a temperature rate of 10°C/min and at 70°C/min, respectively, with the GC Racer are presented in Figure 4.12.

Another approach to achieving fast temperature programming is repackaging a gas chromatographic capillary column in a compartment of low thermal mass (LTM) to provide rapid heating and cooling (RVM Scientific, Inc., www.rvmscientific.com). By replacement of the column door on a gas chromatograph with another door on which the LTM module is installed, fast GC is attainable. In

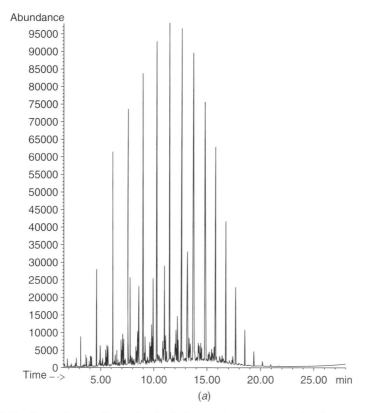

(a)

Figure 4.12 Separations of a diesel fuel sample: (*a*) conventional temperature-programming rate of 10°C/min, 58 cm/s He; (*b*) temperature-programming rate of 70°C/min, 58 cm/s He, all other conditions the same. (Courtesy of Stephen MacDonald.) *(Continued)*

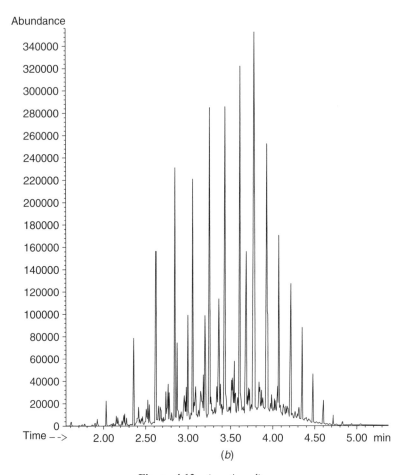

Figure 4.12   (*continued*)

addition, multiple LTM modules can be installed on the same gas chromatograph, permitting independent and simultaneous temperature control for further capability. As is the case with the GC Racer, the same software, inlet systems, detectors, and gas chromatographic column choices can be employed.

4.6   SUBAMBIENT OVEN TEMPERATURE CONTROL

Most gas chromatographs have the capability to operate the column oven at subambient temperatures. An accessory kit is available for either liquid nitrogen ($-99°C$) or carbon dioxide ($-40°C$) as a coolant and includes a cryogenic valve that is microprocessor controlled. The valve opens and closes depending on the demand for coolant. In the open position, coolant is sprayed into the oven, where it chills the oven down with assistance from forced-air convection.

REFERENCES

1. W. E. Harris and H. W. Habgood, *Programmed Temperature Gas Chromatography,* Wiley, New York, 1966.
2. J. V. Hinshaw, in *Modern Practice of Gas Chromatography*, R. L. Grob and E. F. Barry, Eds., Wiley, Hoboken, NJ, 2004, Chap. 4.
3. G. N. Abbay, E. F. Barry, S. Leepipatpiboon, R. Ramstad, M. C. Roman, R. W. Siergiej, L. R. Snyder, and W. L. Winniford, *LC/GC*, 9, 100–114 (1991).
4. D. E. Bautz, J. W. Dolan, W. D. Raddatz, and L. R. Snyder, *Anal. Chem.*, **62**, 1560 (1990).
5. D. E. Bautz, J. W. Dolan, and L. R. Snyder, *J. Chromatogr.*, 541, 1 (1991).
6. L. R. Snyder, D. E. Bautz, and J. W. Dolan, *J. Chromatogr.*, 541, 35 (1991).
7. R. L. Grob, E. F. Barry, S. Leepipatpiboon, J. M. Ombaba, and L. A. Colon, *J. Chromatogr. Sci.*, 30, 177 (1992).
8. R. Sacks, H. Smith, and M. Nowak, *Anal. Chem.*, **70**, 29A (1998).
9. R. Sacks. C. Coutant, and A. Grall, *Anal. Chem.*, **72**, 524A (2000).
10. J. Hinshaw and L. Ettre, *Chromatographia*, **21**, 561 (1986).
11. J. Hinshaw and L. Ettre, *Chromatographia*, **21**, 669 (1986).
12. P. Sandra, F. David, M. Proot, G. Diricks, M. Verstappeand, and M. Verzele, *J. High Resolut. Chromatogr. Commun.*, **8**, 782 (1985).
13. C. Coutant, A. Grall, and R. Sacks, *Anal. Chem.*, **71**, 2123 (1999).
14. A. Grall, C. Coutant, and R. Sacks, *Anal. Chem.*, **72**, 591 (2000).
15. A. van Es, J. Janssen, C. Cramers, and J. Rijks, *J. High Resolut. Chromatogr. Commun.*, **11**, 852 (1988).
16. M. Klemp, M. Akard, and R. Sacks, *Anal. Chem.*, **65**, 2516 (1993).
17. Z. Liu and J. B. Phillips, *J. Microcolumn Sep.*, **1**, 249 (1989).
18. Special issue of *High Resolut. Chromatogr. Chromatogr. Commun.*, **23**(3) (2000).
19. R. Sacks, in *Modern Practice of Gas Chromatography*, R. L. Grob and E. F. Barry, Eds., Wiley, Hoboken, NJ, 2004, Chap. 5.
20. G. M. Gross, B. J. Prazen, J. W. Grate, and R. E. Synovec, *Anal. Chem.*, **76**, 3515 (2004).
21. R. Laub and J. Purnell, *Anal. Chem.*, **48**, 1720 (1976).
22. R. Laub and J. Purnell, *Anal. Chem.*, **48**, 799 (1976).
23. J. Purnell and M. Watten, *J. Chromatogr.*, **555**, 173 (1991).
24. D. Deans and I. Scott, *Anal. Chem.*, **45**, 1137 (1973).
25. H. Smith and R. Sacks, *Anal. Chem.*, **69**, 5159 (1997).
26. M. Akard and R. Sacks, *Anal. Chem.*, **67**, 273 (1995).
27. C. Leonard and R. Sacks, *Anal. Chem.*, **71**, 5501 (1999).
28. T. Veriotti, M. McGuigan, and R. Sacks, *Anal. Chem.*, **73**, 279 (2001).
29. T. Veriotti and R. Sacks, *Anal. Chem.*, **73**, 813 (2001).
30. T. Veriotti and R. Sacks, *Anal. Chem.*, **73**, 3045 (2001).
31. J. Whiting and R. Sacks, *Anal. Chem.*, **74**, 246 (2002).
32. R. E. Wittig, F. L. Dorman, C. M. English, and R. D. Sachs, *J. Chromatogr. A*, **1027**, 75 (2004).
33. J. Harynuk and P. J. Marriott, *Anal. Chem.*, **78**, 2028 (2006).

SELECTED REFERENCES

Below are listed titles of books available which cover the techniques of chromatography in its broadest sense as well as book-titles focused on gas chromatography which contain a treatment of column performance.

Analytical Gas Chromatography, 2nd ed., by W. Jennings, E. Mittlefehldt, and P. Stremple, Academic Press, San Diego, CA, 1997.

Basic Gas Chromatography, by H. M. McNair and J. M. Miller, Wiley, New York, 1997.

Capillary Gas Adsorption Chromatography, by V. G. Berezkin and J. de Zeeuw, Alfred Heuthig Publishers, Heidelberg, Germany, 1996.

Capillary Gas Chromatography, by D. W. Grant, Wiley, Chichester, West Sussex, England, 1996.

Chemical Separations: Principles, Techniques and Experiments, by C. E. Meloan, Wiley, New York, 1999.

Chromatography: Concepts and Contrasts, 2nd ed., by J. M. Miller, Wiley, Hoboken, NJ, 2004.

Chromatography Theory, by J. Cazes and R. P. W. Scott, Taylor & Francis, London, 2002.

Current Practice of Gas Chromatography–Mass Spectrometry, by W. M. A. Niessen, Taylor & Francis, London, 2001.

Essence of Chromatography, by C. F. Poole, Elsevier, New York, 2002.

Gas Chromatography: Analytical Chemistry by Open Learning, 2nd ed., by I. A. Fowlis, Wiley, New York, 1995.

Gas Chromatography: Biochemical, Biomedical, and Clinical Applications, by R. E. Clement (Ed.), Wiley, New York, 1990.

Gas Chromatography and Mass Spectrometry: A Practical Guide, by F. G. Kitson, B. S. Larsen, and C. N. McEwen, Academic Press, San Diego, CA, 1996.

Gas Chromatographic Techniques and Applications, by A. J. Handley and E. R. Adlard, CRC Press, Boca Raton, FL, 2001.

GC/MS: A Practical User's Guide, by M. McMaster and C. McMaster, Wiley, New York, 1998.

Handbook of Analytical Derivatization Reactions, by D. R. Knapp, Wiley, New York, 1979.

Introduction to Analytical Gas Chromatography, 2nd ed., by R. P. W. Scott, Taylor & Francis, London, 1997.

Milestones in the Evolution of Chromatography, by L. S. Ettre, ChromSource, Wiley, New York, 2002.

Modern Practice of Gas Chromatography, 4th ed., by R. L. Grob and E. F. Barry, Wiley-Interscience, Hoboken, NJ, 2004.

A Practical Guide to the Care, Maintenance, and Troubleshooting of Capillary Gas Chromatographic Systems, 3rd ed., by D. Rood, Wiley, New York, 1999.

Practical Introduction to GC-MS Analysis with Quadrupoles, by M. Oehme, Wiley, New York, 1999.

Techniques and Practice of Chromatography, by R. P. W. Scott, Taylor & Francis, London, 1995.

Unified Separation Science, by J. Calvin Giddings, Wiley, New York, 1991.

APPENDIX A
Guide to Selection of Packed Columns

The packed column separations that follow provide effective guidance for selection of the proper stationary phase, support material, tubing, and so on, and to simplify the process. The easiest way to obtain a column is to select one from the following pages. Another approach is to seek guidance from the literature or the Web sites of column manufacturers.

The authors express their gratitude to Russel Gant and Jill Thomas of Supelco for permission and assistance for the reproduction of the separations presented in *Supelco Bulletin 890B.*

Columns for Gas Chromatography: Performance and Selection, by Eugene F. Barry and Robert L. Grob
Copyright © 2007 John Wiley & Sons, Inc.

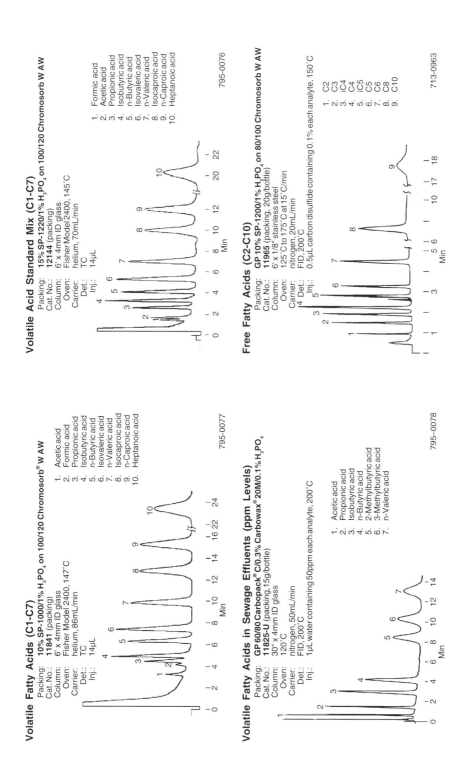

Volatile Fatty Acids (C1-C7)

Packing: 10% SP-1000/1% H_3PO_4 on 100/120 Chromosorb® W AW
Cat. No.: 11841 (packing)
Column: 6' x 4mm ID glass
Oven: Fisher Model 2400, 147°C
Carrier: helium, 86mL/min
Det.: TC
Inj.: 14µL

1. Acetic acid
2. Formic acid
3. Propionic acid
4. Isobutyric acid
5. n-Butyric acid
6. Isovaleric acid
7. n-Valeric acid
8. Isocaproic acid
9. n-Caproic acid
10. Heptanoic acid

795-0077

Volatile Acid Standard Mix (C1-C7)

Packing: 15% SP-1220/1% H_3PO_4 on 100/120 Chromosorb W AW
Cat. No.: 12144 (packing)
Column: 6' x 4mm ID glass
Oven: Fisher Model 2400, 145°C
Carrier: helium, 70mL/min
Det.: TC
Inj.: 14µL

1. Formic acid
2. Acetic acid
3. Propionic acid
4. Isobutyric acid
5. n-Butyric acid
6. Isovaleric acid
7. n-Valeric acid
8. Isocaproic acid
9. n-Caproic acid
10. Heptanoic acid

795-0076

Volatile Fatty Acids in Sewage Effluents (ppm Levels)

Packing: GP 60/80 Carbopack® C/0.3% Carbowax® 20M/0.1% H_3PO_4
Cat. No.: 11825-U (packing, 15g/bottle)
Column: 30" x 4mm ID glass
Oven: 120°C
Carrier: nitrogen, 50mL/min
Det.: FID, 200°C
Inj.: 1µL water containing 50ppm each analyte, 200°C

1. Acetic acid
2. Propionic acid
3. Isobutyric acid
4. n-Butyric acid
5. 2-Methylbutyric acid
6. 3-Methylbutyric acid
7. n-Valeric acid

795-0078

Free Fatty Acids (C2-C10)

Packing: GP 10% SP-1200/1% H_3PO_4 on 80/100 Chromosorb W AW
Cat. No.: 11965 (packing, 20g/bottle)
Column: 6' x 1/8" stainless steel
Oven: 125°C to 175°C at 15°C/min
Carrier: nitrogen, 20mL/min
Det.: FID, 200°C
Inj.: 0.5µL carbon disulfide containing 0.1% each analyte, 150°C

1. C2
2. C3
3. iC4
4. C4
5. iC5
6. C5
7. C6
8. C8
9. C10

713-0963

233

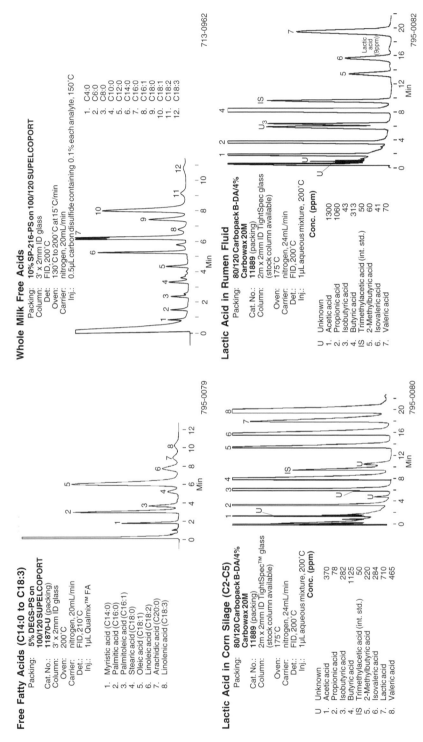

Free Fatty Acids (C14:0 to C18:3)

Packing: **5% DEGS-PS on 100/120 SUPELCOPORT**
Cat. No.: **11870-U** (packing)
Column: 3' x 2mm ID glass
Oven: 200°C
Carrier: nitrogen, 20mL/min
Det.: FID, 210°C
Inj.: 1µL Qualmix™ FA

1. Myristic acid (C14:0)
2. Palmitic acid (C16:0)
3. Palmitoleic acid (C16:1)
4. Stearic acid (C18:0)
5. Oleic acid (C18:1)
6. Linoleic acid (C18:2)
7. Arachidic acid (C20:0)
8. Linolenic acid (C18:3)

795-0079

Whole Milk Free Acids

Packing: **10% SP-216-PS on 100/120 SUPELCOPORT**
Column: 3' x 2mm ID glass
Det.: FID, 200°C
Oven: 130°C to 200°C at 15°C/min
Carrier: nitrogen, 20mL/min
Inj.: 0.5µL carbon disulfide containing 0.1% each analyte, 150°C

1. C4:0
2. C6:0
3. C8:0
4. C10:0
5. C12:0
6. C14:0
7. C16:0
8. C16:1
9. C18:0
10. C18:1
11. C18:2
12. C18:3

713-0962

Lactic Acid in Corn Silage (C2-C5)

Packing: **80/120 Carbopack B-DA/4% Carbowax 20M**
Cat. No.: **11889** (packing)
Column: 2m x 2mm ID TightSpec™ glass (stock column available)
Oven: 175°C
Carrier: nitrogen, 24mL/min
Det.: FID, 200°C
Inj.: 1µL aqueous mixture, 200°C

		Conc. (ppm)
U	Unknown	
1.	Acetic acid	370
2.	Propionic acid	78
3.	Isobutyric acid	282
4.	Butyric acid	1125
IS	Trimethylacetic acid (int. std.)	50
5.	2-Methylbutyric acid	220
6.	Isovaleric acid	284
7.	Lactic acid	710
8.	Valeric acid	465

795-0080

Lactic Acid in Rumen Fluid

Packing: **80/120 Carbopack B-DA/4% Carbowax 20M**
Cat. No.: **11889** (packing)
Column: 2m x 2mm ID TightSpec glass (stock column available)
Oven: 175°C
Carrier: nitrogen, 24mL/min
Det.: FID, 200°C
Inj.: 1µL aqueous mixture, 200°C

		Conc. (ppm)
U	Unknown	
1.	Acetic acid	1300
2.	Propionic acid	1060
3.	Isobutyric acid	43
4.	Butyric acid	313
IS	Trimethylacetic acid (int. std.)	50
5.	2-Methylbutyric acid	60
6.	Isovaleric acid	41
7.	Valeric acid	70

Lactic acid (9ppm)

795-0082

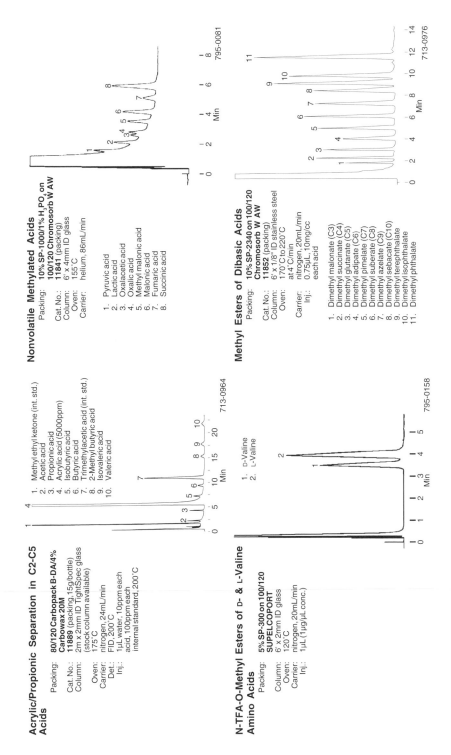

Acrylic/Propionic Separation in C2-C5 Acids

Packing: **80/120 Carbopack B-DA/4% Carbowax 20M**
Cat. No.: **11889** (packing, 15g/bottle)
Column: 2m x 2mm ID TightSpec glass (stock column available)
Oven: 175°C
Carrier: nitrogen, 24mL/min
Det.: FID, 200°C
Inj.: 1μL water, 10ppm each acid, 100ppm each internal standard, 200°C

1. Methyl ethyl ketone (int. std.)
2. Acetic acid
3. Propionic acid
4. Acrylic acid (5000ppm)
5. Isobutyric acid
6. Butyric acid
7. Trimethylacetic acid (int. std.)
8. 2-Methyl butyric acid
9. Isovaleric acid
10. Valeric acid

713-0964

Nonvolatile Methylated Acids

Packing: **10% SP-1000/1% H₃PO₄ on 100/120 Chromosorb W AW**
Cat. No.: **11841** (packing)
Column: 6' x 4mm ID glass
Oven: 155°C
Carrier: helium, 86mL/min

1. Pyruvic acid
2. Lactic acid
3. Oxalacetic acid
4. Oxalic acid
5. Methyl malonic acid
6. Malonic acid
7. Fumaric acid
8. Succinic acid

795-0081

N-TFA-O-Methyl Esters of D- & L-Valine Amino Acids

Packing: **5% SP-300 on 100/120 SUPELCOPORT**
Column: 6' x 2mm ID glass
Oven: 120°C
Carrier: nitrogen, 20mL/min
Inj.: 1μL (1μg/μL conc.)

1. D-Valine
2. L-Valine

795-0158

Methyl Esters of Dibasic Acids

Packing: **10% SP-2340 on 100/120 Chromosorb W AW**
Cat. No.: **11852** (packing)
Column: 6' x 1/8" ID stainless steel
Oven: 170°C to 220°C at 4°C/min
Carrier: nitrogen, 20mL/min
Inj.: 0.75μL, 10mg/cc each acid

1. Dimethyl malonate (C3)
2. Dimethyl succinate (C4)
3. Dimethyl glutarate (C5)
4. Dimethyl adipate (C6)
5. Dimethyl pimelate (C7)
6. Dimethyl suberate (C8)
7. Dimethyl azelate (C9)
8. Dimethyl sebacate (C10)
9. Dimethyl terephthalate
10. Dimethyl isophthalate
11. Dimethyl phthalate

713-0976

235

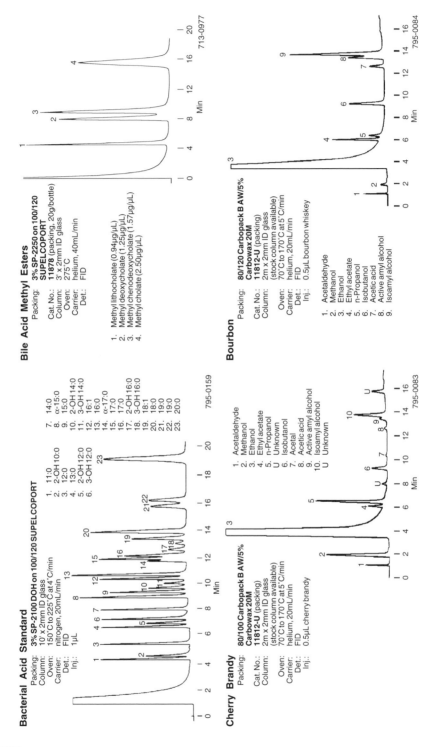

Bacterial Acid Standard

Packing: 3% SP-2100 DOH on 100/120 SUPELCOPORT
Column: 10' x 2mm ID glass
Oven: 150°C to 225°C at 4°C/min
Carrier: nitrogen, 20mL/min
Det.: FID
Inj.: 1µL

1. 11:0
2. 2-OH 10:0
3. 12:0
4. 13:0
5. 2-OH 12:0
6. 3-OH 12:0
7. 14:0
8. α-15:0
9. 15:0
10. 2-OH 14:0
11. 3-OH 14:0
12. 16:1
13. 16:0
14. α-17:0
15. 17:0
16. 17:0
17. 2-OH 16:0
18. 3-OH 16:0
19. 18:1
20. 18:0
21. 19:0
22. 19:0
23. 20:0

795-0159

Bile Acid Methyl Esters

Packing: 3% SP-2250 on 100/120 SUPELCOPORT
Cat. No.: 11878 (packing, 20g/bottle)
Column: 3' x 2mm ID glass
Oven: 275°C
Carrier: helium, 40mL/min
Det.: FID

1. Methyl lithocholate (0.94µg/µL)
2. Methyl deoxycholate (1.25µg/µL)
3. Methyl chenodeoxycholate (1.57µg/µL)
4. Methyl cholate (2.50µg/µL)

713-0977

Cherry Brandy

Packing: 80/100 Carbopack B AW/5% Carbowax 20M
Cat. No.: 11812-U (packing)
Column: 2m x 2mm ID glass
(stock column available)
Oven: 70°C to 170°C at 5°C/min
Carrier: helium, 20mL/min
Det.: FID
Inj.: 0.5µL cherry brandy

1. Acetaldehyde
2. Methanol
3. Ethanol
4. Ethyl acetate
5. n-Propanol
U. Unknown
6. Isobutanol
7. Acetal
8. Acetic acid
9. Active amyl alcohol
10. Isoamyl alcohol
U. Unknown

795-0083

Bourbon

Packing: 80/120 Carbopack B AW/5% Carbowax 20M
Cat. No.: 11812-U (packing)
Column: 2m x 2mm ID glass
(stock column available)
Oven: 70°C to 170°C at 5°C/min
Carrier: helium, 20mL/min
Det.: FID
Inj.: 0.5µL bourbon whiskey

1. Acetaldehyde
2. Methanol
3. Ethanol
4. Ethyl acetate
5. Isobutanol
6. n-Propanol
7. Acetic acid
8. Active amyl alcohol
9. Isoamyl alcohol

795-0084

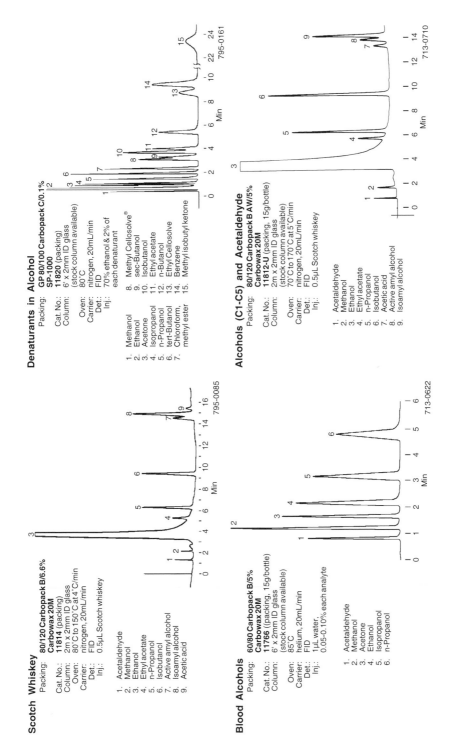

Scotch Whiskey

Packing: **80/120 Carbopack B/6.6%**
Carbowax 20M
Cat. No.: **11814 (packing)**
Column: 2m x 2mm ID glass
Oven: 80°C to 150°C at 4°C/min
Carrier: nitrogen, 20mL/min
Det.: FID
Inj.: 0.5μL Scotch whiskey

1. Acetaldehyde
2. Methanol
3. Ethanol
4. Ethyl acetate
5. n-Propanol
6. Isobutanol
7. Active amyl alcohol
8. Isoamyl alcohol
9. Acetic acid

795-0085

Blood Alcohols

Packing: **60/80 Carbopack B/5%**
Carbowax 20M
Cat. No.: **11766 ((packing, 15g/bottle)**
Column: 6' x 2mm ID glass
(stock column available)
Oven: 85°C
Carrier: helium, 20mL/min
Det.: FID
Inj.: 1μL water,
0.05-0.10% each analyte

1. Acetaldehyde
2. Methanol
3. Acetone
4. Ethanol
5. Isopropanol
6. n-Propanol

713-0622

Denaturants in Alcohol

Packing: **GP 80/100 Carbopack C/0.1%**
SP-1000
Cat. No.: **11820 (packing)**
Column: 6' x 2mm ID glass
(stock column available)
Oven: 80°C
Carrier: nitrogen, 20mL/min
Det.: FID
Inj.: 70% ethanol & 2% of
each denaturant

1. Methanol
2. Ethanol
3. Acetone
4. Isopropanol
5. n-Propanol
6. tert-Butanol
7. Chloroform,
methyl ester
8. Methyl Cellosolve®
9. sec-Butanol
10. Isobutanol
11. Ethyl acetate
12. n-Butanol
13. Ethyl Cellosolve
14. Benzene
15. Methyl isobutyl ketone

795-0161

Alcohols (C1-C5) and Acetaldehyde

Packing: **80/120 Carbopack B AW/5%**
Carbowax 20M
Cat. No.: **11812-U (packing, 15g/bottle)**
Column: 2m x 2mm ID glass
(stock column available)
Oven: 70°C to 170°C at 5°C/min
Carrier: nitrogen, 20mL/min
Det.: FID
Inj.: 0.5μL Scotch whiskey

1. Acetaldehyde
2. Methanol
3. Ethanol
4. Ethyl acetate
5. n-Propanol
6. Isobutanol
7. Acetic acid
8. Active amyl alcohol
9. Isoamyl alcohol

713-0710

237

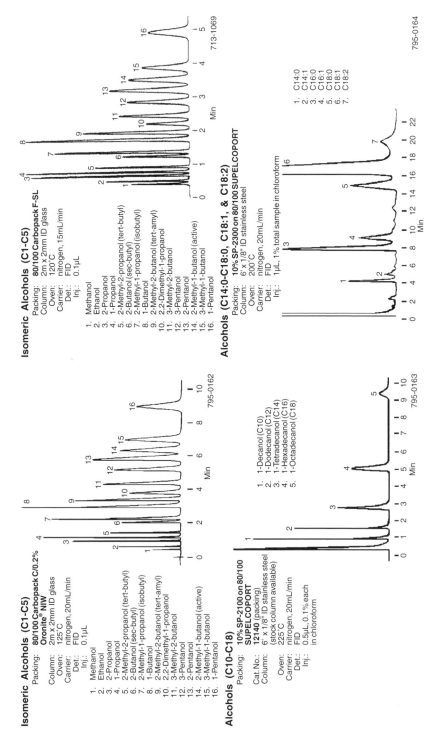

Isomeric Alcohols (C1-C5)

Packing: 80/100 Carbopack C/0.2% Oronite® NIW
Column: 2m x 2mm ID glass
Oven: 125°C
Carrier: nitrogen, 20mL/min
Det.: FID
Inj.: 0.1μL

1. Methanol
2. Ethanol
3. 2-Propanol
4. 1-Propanol
5. 2-Methyl-2-propanol (tert-butyl)
6. 2-Butanol (sec-butyl)
7. 2-Methyl-1-propanol (isobutyl)
8. 1-Butanol
9. 2-Methyl-2-butanol (tert-amyl)
10. 2,2-Dimethyl-1-propanol
11. 3-Methyl-2-butanol
12. 3-Pentanol
13. 2-Pentanol
14. 2-Methyl-1-butanol (active)
15. 3-Methyl-1-butanol
16. 1-Pentanol

795-0162

Isomeric Alcohols (C1-C5)

Packing: 80/100 Carbopack F-SL
Column: 2m x 2mm ID glass
Oven: 120°C
Carrier: nitrogen, 15mL/min
Det.: FID
Inj.: 0.1μL

1. Methanol
2. Ethanol
3. 2-Propanol
4. 1-Propanol
5. 2-Methyl-2-propanol (tert-butyl)
6. 2-Butanol (sec-butyl)
7. 2-Methyl-1-propanol (isobutyl)
8. 1-Butanol
9. 2-Methyl-2-butanol (tert-amyl)
10. 2,2-Dimethyl-1-propanol
11. 3-Methyl-2-butanol
12. 3-Pentanol
13. 2-Pentanol
14. 2-Methyl-1-butanol (active)
15. 3-Methyl-1-butanol
16. 1-Pentanol

713-1069

Alcohols (C10-C18)

Packing: 10% SP-2100 on 80/100 SUPELCOPORT
Cat. No.: 12140 (packing)
Column: 6' x 1/8" ID stainless steel (stock column available)
Oven: 225°C
Carrier: nitrogen, 20mL/min
Det.: FID
Inj.: 0.5μL, 0.1% each in chloroform

1. 1-Decanol (C10)
2. 1-Dodecanol (C12)
3. 1-Tetradecanol (C14)
4. 1-Hexadecanol (C16)
5. 1-Octadecanol (C18)

795-0163

Alcohols (C14:0-C18:0, C18:1, & C18:2)

Packing: 10% SP-2300 on 80/100 SUPELCOPORT
Column: 6' x 1/8" ID stainless steel
Oven: 200°C
Carrier: nitrogen, 20mL/min
Det.: FID
Inj.: 1μL, 1% total sample in chloroform

1. C14:0
2. C14:1
3. C16:0
4. C16:1
5. C18:0
6. C18:1
7. C18:2

795-0164

238

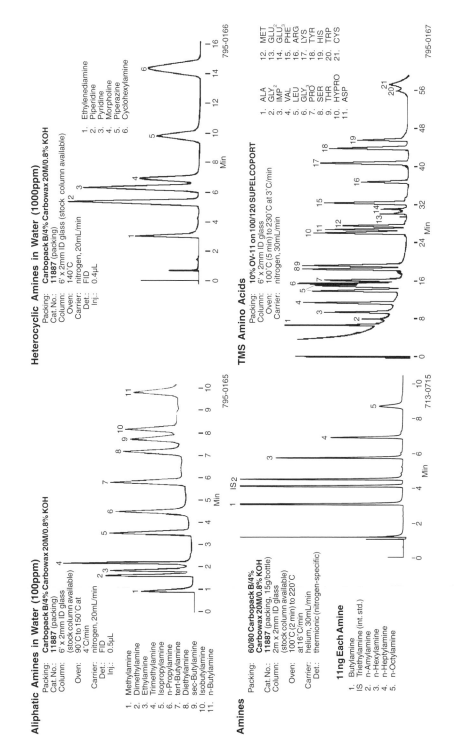

Aliphatic Amines in Water (100ppm)

Packing: **Carbopack B/4% Carbowax 20M/0.8% KOH**
Cat. No.: **11887** (packing)
Column: 6' x 2mm ID glass
        (stock column available)
Oven: 90°C to 150°C at
      4°C/min
Carrier: nitrogen, 20mL/min
Det.: FID
Inj.: 0.5µL

1. Methylamine
2. Dimethylamine
3. Ethylamine
4. Trimethylamine
5. Isopropylamine
6. n-Propylamine
7. tert-Butylamine
8. Diethylamine
9. sec-Butylamine
10. Isobutylamine
11. n-Butylamine

795-0165

Heterocyclic Amines in Water (1000ppm)

Packing: **Carbopack B/4% Carbowax 20M/0.8% KOH**
Cat. No.: **11887** (packing)
Column: 6' x 2mm ID glass (stock column available)
Oven: 140°C
Carrier: nitrogen, 20mL/min
Det.: FID
Inj.: 0.4µL

1. Ethylenediamine
2. Piperidine
3. Pyridine
4. Morpholine
5. Piperazine
6. Cyclohexylamine

795-0166

Amines

Packing: **60/80 Carbopack B/4%**
         Carbowax 20M/0.8% KOH
Cat. No.: **11887** (packing, 15g/bottle)
Column: 2m x 2mm ID glass
        (stock column available)
Oven: 100°C (2 min) to 220°C
      at 16°C/min
Carrier: helium, 30mL/min
Det.: thermionic (nitrogen-specific)

11ng Each Amine

1. Butylamine
IS Triethylamine (int. std.)
2. n-Amylamine
3. n-Hexylamine
4. n-Heptylamine
5. n-Octylamine

713-0715

TMS Amino Acids

Packing: **10% OV-11 on 100/120 SUPELCOPORT**
Column: 6' x 2mm (5 min) ID glass
Oven: 100°C (5 min) to 230°C at 3°C/min
Carrier: nitrogen, 30mL/min

1. ALA
2. GLY^2
3. IMP^2
4. VAL
5. LEU
6. GLY^3
7. PRO^3
8. SER
9. THR
10. HYPRO
11. ASP
12. MET
13. GLU^2
14. GLU^3
15. PHE
16. ARG
17. LYS
18. TYR
19. HIS
20. TRP
21. CYS

795-0167

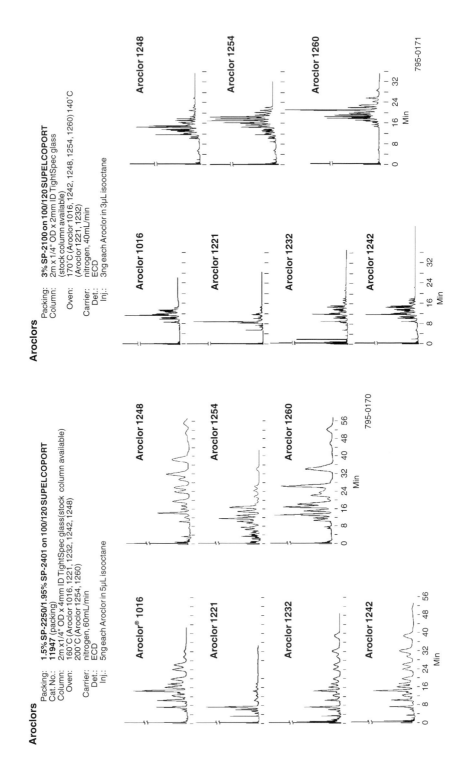

Aroclors

Packing: 1.5% SP-2250/1.95% SP-2401 on 100/120 SUPELCOPORT
Cat. No.: 11947 (packing)
Column: 2m x 1/4" OD x 4mm ID TightSpec glass (stock column available)
Oven: 160°C (Aroclor 1016, 1221, 1232, 1242, 1248)
200°C (Aroclor 1254, 1260)
Carrier: nitrogen, 60mL/min
Det.: ECD
Inj.: 5ng each Aroclor in 5µL isooctane

Aroclor 1248
Aroclor 1254
Aroclor 1260
Aroclor® 1016
Aroclor 1221
Aroclor 1232
Aroclor 1242

795-0170

Aroclors

Packing: 3% SP-2100 on 100/120 SUPELCOPORT
Column: 2m x 1/4" OD x 2mm ID TightSpec glass (stock column available)
Oven: 170°C (Aroclor 1016, 1242, 1248, 1254, 1260) 140°C (Aroclor 1221, 1232)
Carrier: nitrogen, 40mL/min
Det.: ECD
Inj.: 3ng each Aroclor in 3µL isooctane

Aroclor 1248
Aroclor 1254
Aroclor 1260
Aroclor 1016
Aroclor 1221
Aroclor 1232
Aroclor 1242

795-0171

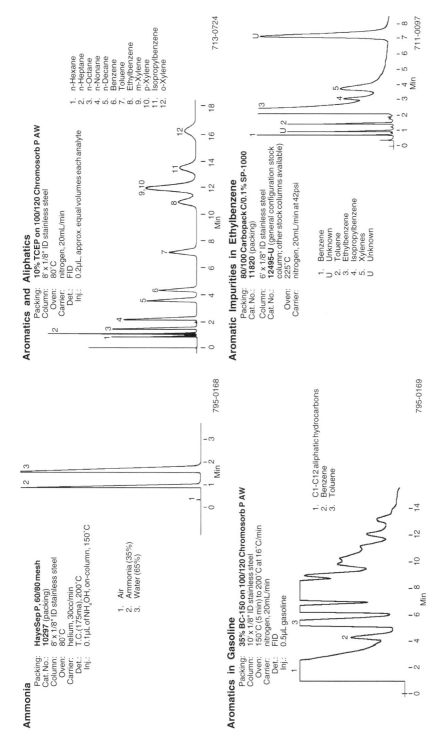

Ammonia

Packing:	**HayeSep P. 60/80 mesh**
Cat. No.:	**10297** (packing)
Column:	8' x 1/8" ID stainless steel
Oven:	80°C
Carrier:	helium, 30cc/min
Det.:	T.C.(175ma), 200°C
Inj.:	0.1µL of NH₄OH, on-column, 150°C

1. Air
2. Ammonia (35%)
3. Water (65%)

795-0168

Aromatics and Aliphatics

Packing:	**10% TCEP on 100/120 Chromosorb P AW**
Column:	8 x 1/8" ID stainless steel
Oven:	80°C
Carrier:	nitrogen, 20mL/min
Det.:	FID
Inj.:	0.2µL, approx. equal volumes each analyte

1. n-Hexane
2. n-Heptane
3. n-Octane
4. n-Nonane
5. n-Decane
6. Benzene
7. Toluene
8. Ethylbenzene
9. m-Xylene
10. p-Xylene
11. Isopropylbenzene
12. o-Xylene

713-0724

Aromatics in Gasoline

Packing:	**35% BC-150 on 100/120 Chromosorb P AW**
Column:	10'x 1/8" ID stainless steel
Oven:	150°C (5 min) to 200°C at 16°C/min
Carrier:	nitrogen, 20mL/min
Det.:	FID
Inj.:	0.5µL gasoline

1. C1-C12 aliphatic hydrocarbons
2. Benzene
3. Toluene

795-0169

Aromatic Impurities in Ethylbenzene

Packing:	**80/100 Carbopack C/0.1% SP-1000**
Cat. No.:	**11820** (packing)
Column:	6' x 1/8" ID stainless steel
Cat. No.:	**12495-U** (general configuration stock column; other stock columns available)
Oven:	225°C
Carrier:	nitrogen, 20mL/min at 42psi

1. Benzene
U. Unknown
2. Toluene
3. Ethylbenzene
4. Isopropylbenzene
5. Xylenes
U. Unknown

711-0097

241

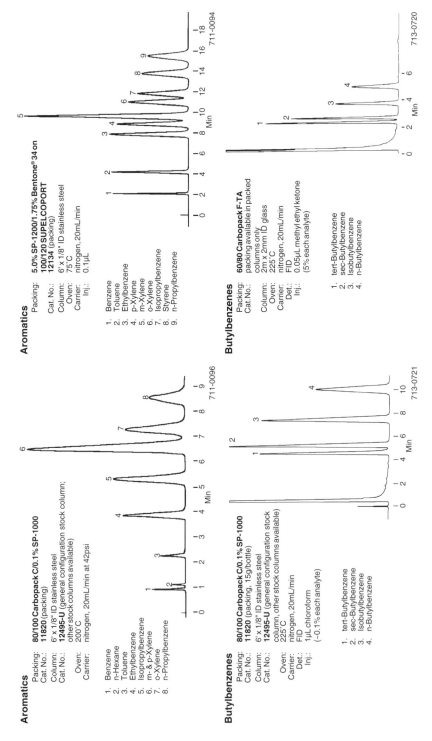

Aromatics

Packing: **80/100 Carbopack C/0.1% SP-1000**
Cat. No.: **11820** (packing)
Column: 6' x 1/8" ID stainless steel
Cat. No.: **12495-U** (general configuration stock column; other stock columns available)
Oven: 200°C
Carrier: nitrogen, 20mL/min at 42psi

1. Benzene
2. n-Hexane
3. Toluene
4. Ethylbenzene
5. Isopropylbenzene
6. m- & p-Xylene
7. o-Xylene
8. n-Propylbenzene

711-0096

Butylbenzenes

Packing: **80/100 Carbopack C/0.1% SP-1000**
Cat. No.: **11820** (packing, 15g/bottle)
Column: 6' x 1/8" ID stainless steel
Cat. No.: **12495-U** (general configuration stock column, other stock columns available)
Oven: 225°C
Carrier: nitrogen, 20mL/min
Det.: FID
Inj.: 1µL chloroform
(~0.1% each analyte)

1. tert-Butylbenzene
2. sec-Butylbenzene
3. Isobutylbenzene
4. n-Butylbenzene

713-0721

Aromatics

Packing: **5.0% SP-1200/1.75% Bentone® 34 on 100/120 SUPELCOPORT**
Cat. No.: **12134** (packing)
Column: 6' x 1/8" ID stainless steel
Oven: 75°C
Carrier: nitrogen, 20mL/min
Inj.: 0.1µL

1. Benzene
2. Toluene
3. Ethylbenzene
4. p-Xylene
5. m-Xylene
6. o-Xylene
7. Isopropylbenzene
8. Styrene
9. n-Propylbenzene

711-0094

Butylbenzenes

Packing: **60/80 Carbopack F-TA**
Cat. No.: packing available in packed columns only
Column: 2m x 2mm ID glass
Oven: 225°C
Carrier: nitrogen, 20mL/min
Det.: FID
Inj.: 0.05µL methyl ethyl ketone (5% each analyte)

1. tert-Butylbenzene
2. sec-Butylbenzene
3. Isobutylbenzene
4. n-Butylbenzene

713-0720

242

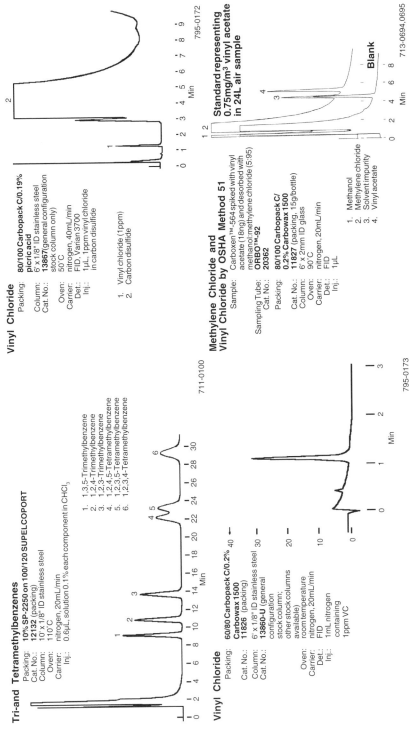

Tri- and Tetramethylbenzenes

Packing: 10% SP-2250 on 100/120 SUPELCOPORT
Cat. No.: 12132 (packing)
Column: 10' x 1/8" ID stainless steel
Oven: 110°C
Carrier: nitrogen, 20mL/min
Inj.: 0.6µL, solution 0.1% each component in CHCl₃

1. 1,3,5-Trimethylbenzene
2. 1,2,4-Trimethylbenzene
3. 1,2,3-Trimethylbenzene
4. 1,2,4,5-Tetramethylbenzene
5. 1,2,3,5-Tetramethylbenzene
6. 1,2,3,4-Tetramethylbenzene

Vinyl Chloride

Packing: 60/80 Carbopack C/0.2%
Carbowax 1500
11826 (packing)
Column: 6' x 1/8" ID stainless steel
Cat. No.: 13860-U (general
configuration
stock column;
other stock columns
available)
Oven: room temperature
Carrier: nitrogen, 20mL/min
Det.: FID
Inj.: 1mL nitrogen
containing
1ppm VC

711-0100

Vinyl Chloride

Packing: 80/100 Carbopack C/0.19%
picric acid
Column: 6' x 1/8" ID stainless steel
Cat. No.: 13867 (general configuration
stock column only)
Oven: 50°C
Carrier: nitrogen, 40mL/min
Det.: FID, Varian 3700
Inj.: 1µL, 1ppm vinyl chloride
in carbon disulfide

1. Vinyl chloride (1ppm)
2. Carbon disulfide

795-0172

Methylene Chloride and
Vinyl Chloride by OSHA Method 51

Sample: Carboxen™-564 spiked with vinyl
acetate (18ng) and desorbed with
methanol:methylene chloride (5:95)
Sampling Tube: ORBO™-92
Cat. No.: 20362
Packing: 80/100 Carbopack C/
0.2% Carbowax 1500
Cat. No.: 11827 (packing, 15g/bottle)
Column: 6' x 2mm ID glass
Oven: 90°C
Carrier: nitrogen, 20mL/min
Det.: FID
Inj.: 1µL

1. Methanol
2. Methylene chloride
3. Solvent impurity
4. Vinyl acetate

**Standard representing
0.75mg/m³ vinyl acetate
in 24L air sample**

Blank

713-0694,0695

795-0173

243

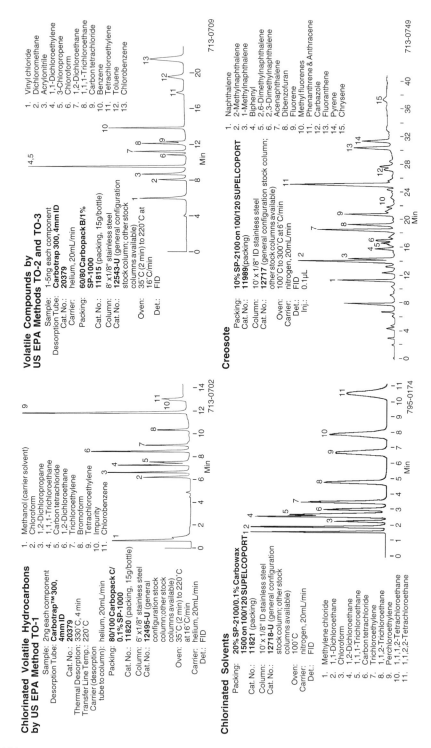

Chlorinated Volatile Hydrocarbons by US EPA Method TO-1

Sample: 2ng each component
Desorption Tube: **Carbotrap™ 300, 4mm ID**
Cat. No.: **20379**
Thermal Desorption: 330°C, 4 min
Transfer Line Temp.: 220°C
Carrier (desorption tube to column): helium, 20mL/min
Packing: **80/100 Carbopack C/ 0.1% SP-1000**
Cat. No.: **11820** (packing, 15g/bottle)
Column: 6′x1/8″ stainless steel
Cat. No.: **12495-U** (general configuration stock column; other stock columns available)
Oven: 35°C (2 min) to 220°C at 16°C/min
Carrier: helium, 20mL/min
Det.: FID

1. Methanol (carrier solvent)
2. Chloroform
3. 1,2-Dichloropropane
4. 1,1,1-Trichloroethane
5. Carbon tetrachloride
6. 1,2-Dichloroethane
7. Trichloroethylene
8. Bromoform
9. Tetrachloroethylene
10. Impurity
11. Chlorobenzene

713-0702

Volatile Compounds by US EPA Methods TO-2 and TO-3

Sample: 1-5ng each component
Desorption Tube: **Carbotrap 300, 4mm ID**
Cat. No.: **20379**
Carrier: helium, 20mL/min
Packing: **60/80 Carbopack B/1% SP-1000**
Cat. No.: **11815** (packing, 15g/bottle)
Column: 8′ x1/8″ stainless steel
Cat. No.: **12543-U** (general configuration stock column; other stock columns available)
Oven: 35°C (2 min) to 220°C at 16°C/min
Det.: FID

1. Vinyl chloride
2. Dichloromethane
3. Acrylonitrile
4. 1,1-Dichloroethylene
5. 3-Chloropropene
6. Chloroform
7. 1,2-Dichloroethane
8. 1,1,1-Trichloroethane
9. Carbon tetrachloride
10. Benzene
11. Tetrachloroethylene
12. Toluene
13. Chlorobenzene

713-0709

Chlorinated Solvents

Packing: **20% SP-2100/0.1% Carbowax 1500 on 100/120 SUPELCOPORT**
Cat. No.: **11821** (packing)
Column: 10′ x 1/8″ ID stainless steel
Cat. No.: **12718-U** (general configuration stock column; other stock columns available)
Oven: 100°C
Carrier: nitrogen, 20mL/min
Det.: FID

1. Methylene chloride
2. 1,1-Dichloroethane
3. Chloroform
4. 1,2-Dichloroethane
5. 1,1,1-Trichloroethane
6. Carbon tetrachloride
7. Trichloroethylene
8. 1,1,2-Trichloroethane
9. Perchloroethylene
10. 1,1,1,2-Tetrachloroethane
11. 1,1,2,2-Tetrachloroethane

795-0174

Creosote

Packing: **10% SP-2100 on 100/120 SUPELCOPORT 11989** (packing)
Cat. No.:
Column: 10′ x 1/8″ ID stainless steel
Cat. No.: **12717** (general configuration stock column; other stock columns available)
Oven: 100°C to 300°C at 6°C/min
Carrier: nitrogen, 20mL/min
Det.: FID
Inj.: 0.1µL

1. Naphthalene
2. 2-Methylnaphthalene
3. 1-Methylnaphthalene
4. Biphenyl
5. 2,6-Dimethylnaphthalene
6. 2,3-Dimethylnaphthalene
7. Acenaphthalene
8. Dibenzofuran
9. Fluorene
10. Methylfluorenes
11. Phenanthrene & Anthracene
12. Carbazole
13. Fluoranthene
14. Pyrene
15. Chrysene

713-0749

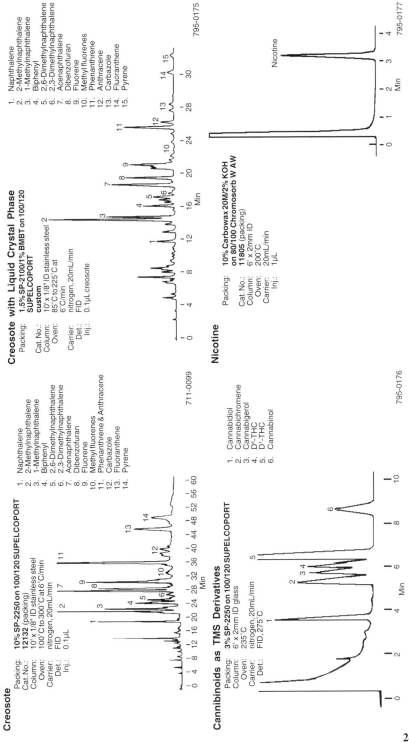

Creosote

Packing: 10% SP-2250 on 100/120 SUPELCOPORT
Cat.No.: 12132 (packing)
Column: 10' x 1/8" ID stainless steel
Oven: 100°C to 300°C at 6°C/min
Carrier: nitrogen, 20mL/min
Det.: FID
Inj.: 0.1µL

1. Naphthalene
2. 2-Methylnaphthalene
3. 1-Methylnaphthalene
4. Biphenyl
5. 2,6-Dimethylnaphthalene
6. 2,3-Dimethylnaphthalene
7. Acenaphthalene
8. Dibenzofuran
9. Fluorene
10. Methylfluorenes
11. Phenanthrene & Anthracene
12. Carbazole
13. Fluoranthene
14. Pyrene

711-0099

Creosote with Liquid Crystal Phase

Packing: 1.5% SP-2100/1% BMBT on 100/120
SUPELCOPORT
custom
Cat.No.: custom
Column: 10' x 1/8" ID stainless steel
Oven: 85°C to 225°C at
6°C/min
Carrier: nitrogen, 20mL/min
Det.: FID
Inj.: 0.1µL creosote

1. Naphthalene
2. 2-Methylnaphthalene
3. 1-Methylnaphthalene
4. Biphenyl
5. 2,6-Dimethylnaphthalene
6. 2,3-Dimethylnaphthalene
7. Acenaphthalene
8. Dibenzofuran
9. Fluorene
10. Methylfluorenes
11. Phenanthrene
12. Anthracene
13. Carbazole
14. Fluoranthene
15. Pyrene

795-0175

Cannibinoids as TMS Derivatives

Packing: 3% SP-2250 on 100/120 SUPELCOPORT
Column: 6' x 2mm ID glass
Oven: 235°C
Carrier: nitrogen, 20mL/min
Det.: FID, 275°C

1. Cannabidiol
2. Cannabichromene
3. Cannabigerol
4. D⁹-THC
5. D¹-THC
6. Cannabinol

795-0176

Nicotine

Packing: 10% Carbowax 20M/2% KOH
on 80/100 Chromosorb W AW
Cat.No.: 11805 (packing)
Column: 6' x 2mm ID
Oven: 200°C
Carrier: 20mL/min
Inj.: 1µL

Nicotine

795-0177

245

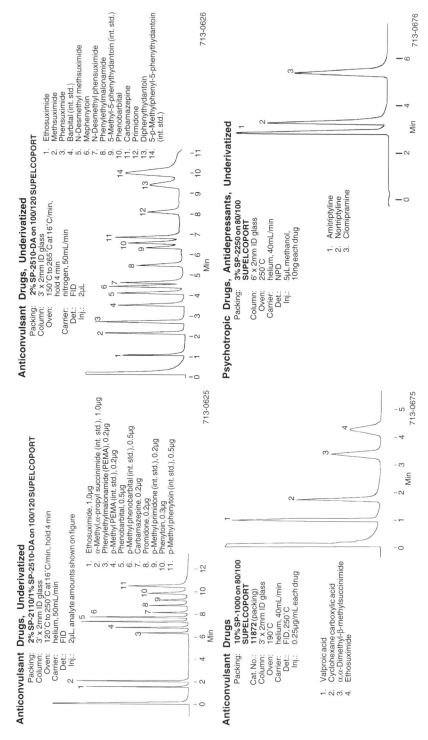

Anticonvulsant Drugs, Underivatized

Packing: **2% SP-2110/1% SP-2510-DA on 100/120 SUPELCOPORT**
Column: 3' x 2mm ID glass
Oven: 120°C to 250°C at 16°C/min, hold 4 min
Carrier: helium, 50mL/min
Det.: FID
Inj.: 2µL, analyte amounts shown on figure

1. Ethosuximide, 1.0µg
2. α-Methyl,α-propyl succinimide (int. std.), 1.0µg
3. Phenylethylmalonamide (PEMA), 0.2µg
4. p-Methyl PEMA (int. std.), 0.2µg
5. Phenobarbital, 0.5µg
6. p-Methyl phenobarbital (int. std.), 0.5µg
7. Carbamazepine, 0.2µg
8. Promidone, 0.2µg
9. p-Methyl primidone (int. std.), 0.2µg
10. Phenytoin, 0.3µg
11. p-Methyl phenytoin (int. std.), 0.5µg

713-0625

Anticonvulsant Drugs

Packing: **10% SP-1000 on 80/100 SUPELCOPORT**
Cat. No.: **11872 (packing)**
Column: 3' x 2mm ID glass
Oven: 190°C
Carrier: helium, 40mL/min
Det.: FID, 250°C
Inj.: 0.25µg/mL each drug

1. Valproic acid
2. Cyclohexane carboxylic acid
3. α,α-Dimethyl-β-methylsuccinimide
4. Ethosuximide

713-0675

Anticonvulsant Drugs, Underivatized

Packing: **2% SP-2510-DA on 100/120 SUPELCOPORT**
Column: 3' x 2mm ID glass
Oven: 150°C to 265°C at 16°C/min, hold 4 min
Carrier: nitrogen, 50mL/min
Det.: FID
Inj.: 2µL

1. Ethosuximide
2. Methsuximide
3. Phensuximide
4. Barbital (int. std.)
5. N-Desmethyl methsuximide
6. Mephenytoin
7. N-Desmethyl phensuximide
8. Phenylethylmalonamide
9. 5-Methyl-5-phenylhydantoin (int. std.)
10. Phenobarbital
11. Carbamazepine
12. Primidone
13. Diphenythydantoin
14. 5-p-Methylphenyl-5-phenythydantoin (int. std.)

713-0626

Psychotropic Drugs, Antidepressants, Underivatized

Packing: **3% SP-2250 on 80/100 SUPELCOPORT**
Column: 6' x 2mm ID glass
Oven: 250°C
Carrier: helium, 40mL/min
Det.: NPD
Inj.: 5µL methanol, 10ng each drug

1. Amitriptyline
2. Nortriptyline
3. Clomipramine

713-0676

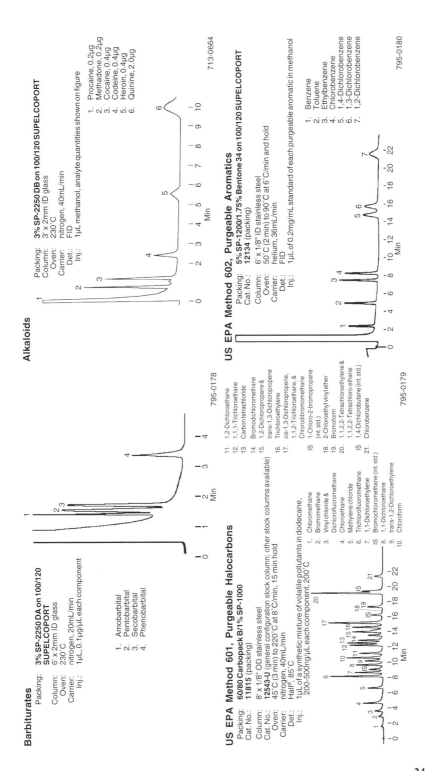

Barbiturates

Packing: 3% SP-2250 DA on 100/120 SUPELCOPORT
Column: 6' x 2mm ID glass
Oven: 230°C
Carrier: nitrogen, 20mL/min
Inj.: 1μL, 0.1μg/μL each component

1. Amobarbital
2. Pentobarbital
3. Secobarbital
4. Phenobarbital

795-0178

Alkaloids

Packing: 3% SP-2250 DB on 100/120 SUPELCOPORT
Column: 3' x 2mm ID glass
Oven: 230°C
Carrier: nitrogen, 40mL/min
Det.: FID
Inj.: 1μL methanol, analyte quantities shown on figure

1. Procaine, 0.2μg
2. Methadone, 0.2μg
3. Cocaine, 0.4μg
4. Codeine, 0.4μg
5. Heroin, 0.4μg
6. Quinine, 2.0μg

713-0664

US EPA Method 601, Purgeable Halocarbons

Packing: 60/80 Carbopack B/1% SP-1000
Cat. No.: 11815 (packing)
Column: 8' x 1/8" OD stainless steel
Cat. No.: 12543-U (general configuration stock column; other stock columns available)
Oven: 45°C (3 min) to 220°C at 8°C/min, 15 min hold
Carrier: nitrogen, 40mL/min
Det.: Hall®, 85°C
Inj.: 1μL of a synthetic mixture of volatile pollutants in dodecane, 200-500ng/μL each component, 200°C

1. Chloromethane
2. Bromomethane
3. Vinyl chloride &
4. Chloroethane
5. Methylene chloride
6. Trichlorofluoromethane
7. 1,1-Dichloroethylene
IS. Bromochloromethane (int. std.)
8. 1,1-Dichloroethane
9. trans-1,2-Dichloroethylene
10. Chloroform

11. 1,2-Dichloroethane
12. 1,1,1-Trichloroethane
13. Carbon tetrachloride
14. Bromodichloromethane
15. 1,2-Dichloropropane &
    trans-1,3-Dichloropropene
16. Trichloroethylene
17. cis-1,3-Dichloropropene,
    1,1,2-Trichloroethane, &
    Chlorodibromomethane
IS. 1-Chloro-2-bromopropane
    (int. std.)
18. 2-Chloroethyl vinyl ether
19. Bromoform
20. 1,1,2,2-Tetrachloroethylene &
    1,1,2,2-Tetrachloro-ethane
IS. 1,4-Dichlorobutane (int. std.)
21. Chlorobenzene

795-0179

US EPA Method 602, Purgeable Aromatics

Packing: 5% SP-1200/1.75% Bentone 34 on 100/120 SUPELCOPORT
Cat. No.: 12134 (packing)
Column: 6' x 1/8" ID stainless steel
Oven: 50°C (2 min) to 90°C at 6°C/min and hold
Carrier: helium, 36mL/min
Det.: FID
Inj.: 1μL of 0.2mg/mL standard of each purgeable aromatic in methanol

1. Benzene
2. Toluene
3. Ethylbenzene
4. Chlorobenzene
5. 1,4-Dichlorobenzene
6. 1,3-Dichlorobenzene
7. 1,2-Dichlorobenzene

795-0180

247

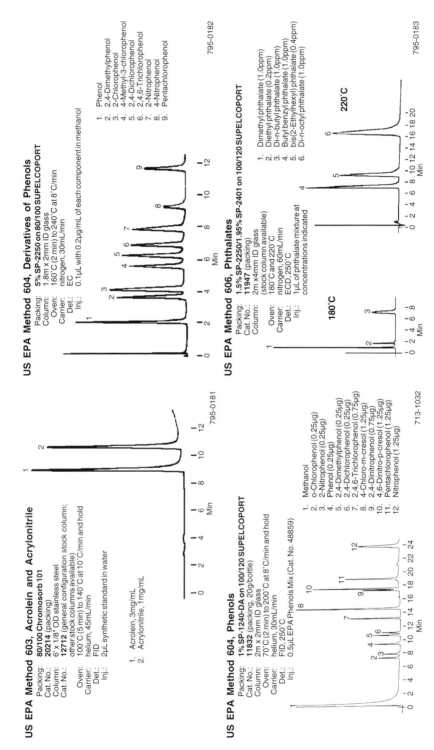

US EPA Method 603, Acrolein and Acrylonitrile

Packing: 80/100 Chromosorb 101
Cat.No.: 20214 (packing)
Column: 6' x 1/8" OD stainless steel
Cat.No.: 12712 (general configuration stock column; other stock columns available)
Oven: 100°C (5 min) to 140°C/min and hold
Carrier: helium, 45mL/min
Det.: FID
Inj.: 2μL synthetic standard in water

1. Acrolein, 3mg/mL
2. Acrylonitrile, 1mg/mL

795-0181

US EPA Method 604, Phenols

Packing: 1% SP-1240-DA on 100/120 SUPELCOPORT
Cat.No.: 11832 (packing, 20g/bottle)
Column: 2m x 2mm ID glass
Oven: 70°C (2 min) to 200°C at 8°C/min and hold
Carrier: helium, 30mL/min
Det.: FID, 250°C
Inj.: 0.5μL EPA Phenols Mix (Cat. No. 48859)

1. Methanol
2. o-Chlorophenol (0.25μg)
3. 2-Nitrophenol (0.25μg)
4. Phenol (0.25μg)
5. 2,4-Dimethylphenol (0.25μg)
6. 2,4-Dichlorophenol (0.25μg)
7. 2,4,6-Trichlorophenol (0.75μg)
8. 4-Chloro-m-cresol (1.25μg)
9. 2,4-Dinitrophenol (0.75μg)
10. 4,6-Dinitro-p-cresol (1.25μg)
11. Pentachlorophenol (1.25μg)
12. Nitrophenol (1.25μg)

713-1032

US EPA Method 604, Derivatives of Phenols

Packing: 5% SP-2250 on 80/100 SUPELCOPORT
Column: 1.8m x 2mm ID glass
Oven: 160°C (2 min) to 240°C at 8°C/min
Carrier: nitrogen, 30mL/min
Det.: EC
Inj.: 0.1μL with 0.2μg/mL of each component in methanol

1. Phenol
2. 2,4-Dimethylphenol
3. 2-Chlorophenol
4. 4-Methyl-3-chlorophenol
5. 2,4-Dichlorophenol
6. 2,4,6-Trichlorophenol
7. 2-Nitrophenol
8. 4-Nitrophenol
9. Pentachlorophenol

795-0182

US EPA Method 606, Phthalates

Packing: 1.5% SP-2250/1.95% SP-2401 on 100/120 SUPELCOPORT
Cat. No.: 11947 (packing)
Column: 2m x4mm ID glass
(stock column available)
Oven: 180°C and 220°C
Carrier: nitrogen, 60mL/min
Det.: ECD, 250°C
Inj.: 1μL of phthalate mixture at concentrations indicated

1. Dimethyl phthalate (1.0ppm)
2. Diethyl phthalate (0.2ppm)
3. Di-n-butyl phthalate (1.0ppm)
4. Butyl benzyl phthalate (1.0ppm)
5. bis(2-Ethylhexyl) phthalate (0.4ppm)
6. Di-n-octyl phthalate (1.0ppm)

795-0183

248

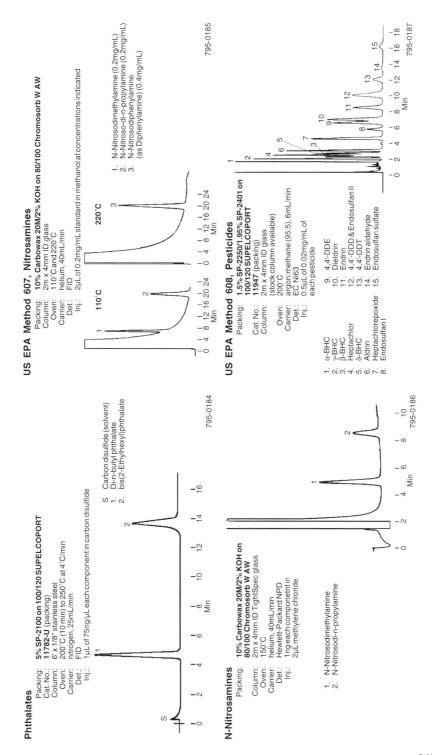

Phthalates

Packing: **5% SP-2100 on 100/120 SUPELCOPORT**
Cat. No.: **11782-U** (packing)
Column: 6' x 1/8" stainless steel
Oven: 200°C (10 min) to 250°C at 4°C/min
Carrier: nitrogen, 25mL/min
Det.: FID
Inj.: 1µL of 75ng/µL each component in carbon disulfide

S. Carbon disulfide (solvent)
1. Di-n-butyl phthalate
2. bis(2-Ethylhexyl)phthalate

795-0184

N-Nitrosamines

Packing: **10% Carbowax 20M/2% KOH on 80/100 Chromosorb W AW**
Column: 2m x 4mm ID TightSpec glass
Oven: 150°C
Carrier: helium, 40mL/min
Det.: Hewlett-Packard NPD
Inj.: 1ng each componetnt in 2µL methylene chloride

1. N-Nitrosodimethylamine
2. N-Nitrosodi-n-propylamine

795-0186

US EPA Method 607, Nitrosamines

Packing: **10% Carbowax 20M/2% KOH on 80/100 Chromosorb W AW**
Column: 2m x 4mm ID glass
Oven: 110°C and 220°C
Carrier: helium, 40mL/min
Det.: FID
Inj.: 2µL of 0.2mg/mL standard in methanol at concentrations indicated

1. N-Nitrosodimethylamine (0.2mg/mL)
2. N-Nitroso-di-n-propylamine (0.2mg/mL)
3. N-Nitrosodiphenylamine (as Diphenylamine) (0.4mg/mL)

795-0185

US EPA Method 608, Pesticides

Packing: **1.5% SP-2250/1.95% SP-2401 on 100/120 SUPELCOPORT**
Cat. No.: **11947** (packing)
Column: 2m x 4mm ID glass (stock column available)
Oven: 200°C
Carrier: argon:methane (95:5), 6mL/min
Det.: EC Ni63
Inj.: 0.5µL of 0.02mg/mL of each pesticide

1. α-BHC
2. γ-BHC
3. β-BHC
4. Heptachlor
5. δ-BHC
6. Aldrin
7. Heptachlorepoxide
8. Endosulfan I
9. 4,4'-DDE
10. Dieldrin
11. Endrin
12. 4,4'-DDD & Endosulfan II
13. 4,4'-DDT
14. Endrin aldehyde
15. Endosulfan sulfate

795-0187

249

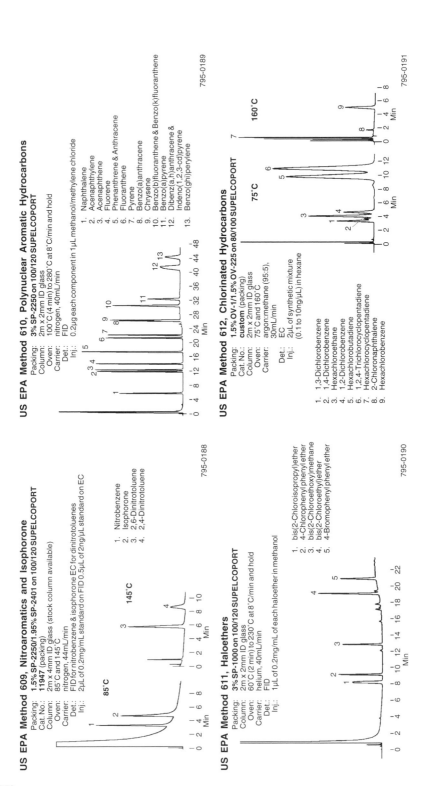

US EPA Method 609, Nitroaromatics and Isophorone

Packing: 1.5% SP-2250/1.95% SP-2401 on 100/120 SUPELCOPORT
Cat. No.: 11947 (packing)
Column: 2m x 4mm ID glass (stock column available)
Oven: 85°C and 145°C
Carrier: nitrogen, 44mL/min
Det.: FID for nitrobenzene & isophorone EC for dinitrotoluenes
Inj.: 2µL of 0.2mg/mL standard on FID 0.5µL of 2ng/µL standard on EC

1. Nitrobenzene
2. Isophorone
3. 2,6-Dinitrotoluene
4. 2,4-Dinitrotoluene

145°C
85°C

795-0188

US EPA Method 611, Haloethers

Packing: 3% SP-1000 on 100/120 SUPELCOPORT
Column: 2m x 2mm ID glass
Oven: 60°C (2 min) to 230°C at 8°C/min and hold
Carrier: helium, 40mL/min
Det.: FID
Inj.: 1µL of 0.2mg/mL of each haloether in methanol

1. bis(2-Chloroisopropyl)ether
2. 4-Chlorophenyl phenyl ether
3. bis(2-Chloroethoxy)methane
4. bis(2-Chloroethyl)ether
5. 4-Bromophenyl phenyl ether

795-0190

US EPA Method 610, Polynuclear Aromatic Hydrocarbons

Packing: 3% SP-2250 on 100/120 SUPELCOPORT
Column: 2m x 2mm ID glass
Oven: 100°C (4 min) to 280°C at 8°C/min and hold
Carrier: nitrogen, 40mL/min
Det.: FID
Inj.: 0.2µg each component in 1µL methanol/methylene chloride

1. Naphthalene
2. Acenaphthylene
3. Acenaphthene
4. Fluorene
5. Phenanthrene & Anthracene
6. Fluoranthene
7. Pyrene
8. Benzo(a)anthracene
9. Chrysene
10. Benzo(b)fluoranthene & Benzo(k)fluoranthene
11. Benzo(a)pyrene
12. Dibenz(a,h)anthracene & Indeno(1,2,3-cd)pyrene
13. Benzo(ghi)perylene

795-0189

US EPA Method 612, Chlorinated Hydrocarbons

Packing: 1.5% OV-1/1.5% OV-225 on 80/100 SUPELCOPORT
Cat. No.: custom (packing)
Column: 2m x 2mm ID glass
Oven: 75°C and 160°C
Carrier: argon:methane (95:5), 30mL/min
Det.: EC
Inj.: 2µL of synthetic mixture (0.1 to 10ng/µL) in hexane

1. 1,3-Dichlorobenzene
2. 1,4-Dichlorobenzene
3. Hexachloroethane
4. 1,2-Dichlorobenzene
5. Hexachlorobutadiene
6. 1,2,4-Trichlorocyclopentadiene
7. Hexachlorocyclopentadiene
8. 2-Chloronaphthalene
9. Hexachlorobenzene

160°C
75°C

795-0191

250

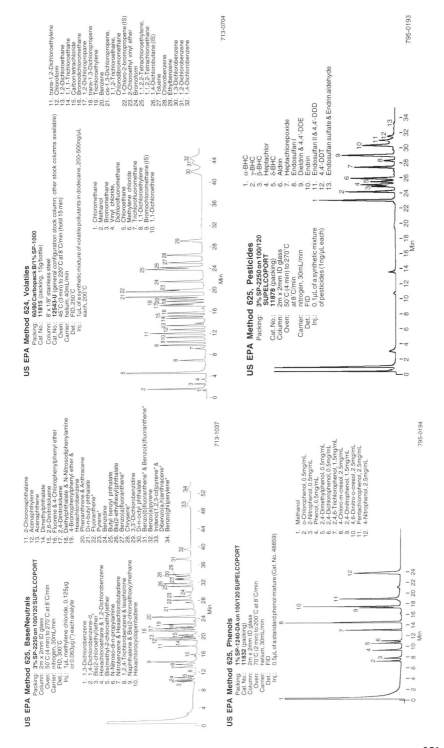

US EPA Method 625, Base/Neutrals

Packing: 3% SP-2250 on 100/120 SUPELCOPORT
Cat. No.:
Column: 2m x 2mm ID glass
Oven: 50°C (4 min) to 270°C at 8°C/min
Det.: FID, 300°C
Carrier: nitrogen, 30mL/min
Inj.: 1μL methylene chloride, 0.125μg
or 0.063μg (*) each analyte

1. 1,3-Dichlorobenzene
2. 1,4-Dichlorobenzene & 1,2-Dichlorobenzene
3. Bis(2-chloroethyl)ether
4. Hexachloroethane & 1,2-Dichlorobenzene
5. Bis(methyl)-2-chloroethyl)ether
6. N-Nitroso-di-n-propylamine
7. Nitrobenzene & Hexachlorobutadiene
8. 1,2,4-Trichlorobenzene & Isophorone
9. Naphthalene & Bis(2-chloroethoxy)methane
10. Hexachlorocyclopentadiene
11. 2-Chloronaphthalene
12. Acenaphthylene
13. Acenaphthene
14. Dimethylphthalate
15. 2,6-Dinitrotoluene
16. Fluorene & 4-Chlorophenylphenyl ether
17. 2,4-Dinitrotoluene
18. Diethylphthalate & N-Nitrosodiphenylamine
19. 4-Bromophenylphenyl ether & Hexachlorobenzene
20. Phenanthrene & Anthracene
21. Di-n-butyl phthalate
22. Fluoranthene*
23. Pyrene*
24. Benzidine
25. Butyl benzyl phthalate
26. Bis(2-ethylhexyl)phthalate
27. Benzo(a)fluoranthene*
28. Chrysene*
29. 3,3-Dichlorobenzidine
30. Di-n-octyl phthalate
31. Benzo(b)fluoranthene* & Benzo(k)fluoranthene*
32. Benzo(a)pyrene
33. Indeno(1,2,3-c,d)pyrene* & Dibenzo(a,h)anthracene*
34. Benzo(ghi)perylene*

713-1037

US EPA Method 625, Phenols

Packing: 1% SP-1240-DA on 100/120 SUPELCOPORT
Cat. No.: 11132 (packing)
Column: 2m x 2mm ID glass
Oven: 70°C (2 min) to 200°C at 8°C/min
Carrier: helium, 30mL/min
Det.: FID
Inj.: 0.5μL of a standard phenol mixture (Cat. No. 48859)

1. Methanol
2. o-Chlorophenol, 0.5mg/mL
3. 2-Nitrophenol, 0.5mg/mL
4. Phenol, 0.5mg/mL
5. 2,4-Dimethylphenol, 0.5mg/mL
6. 2,4-Dichlorophenol, 0.5mg/mL
7. 2,4,6-Trichlorophenol, 1.5mg/mL
8. 4-Chloro-m-cresol, 2.5mg/mL
9. 2,4-Dinitrophenol, 1.5mg/mL
10. 4,6-Dinitro-o-cresol, 2.5mg/mL
11. Pentachlorophenol, 2.5mg/mL
12. 4-Nitrophenol, 2.5mg/mL

795-0194

US EPA Method 624, Volatiles

Packing: 60/80 Carbopack B/1% SP-1000
Cat. No.: 11815 (packing, 15g/bottle)
Column: 8' x 1/8" stainless steel
Cat. No.: 12543-U (as a configuration stock column; other stock columns available)
Oven: 45°C (3 min) to 220°C at 8°C/min (hold 15 min)
Carrier: helium, 40mL/min
Det.: FID, 250°C
Inj.: 1μL of a synthetic mixture of volatile pollutants in dodecane, 200-500ng/μL each, 200°C

1. Chloromethane
2. Methanol
3. Bromomethane
4. Vinyl chloride,
   Dichlorofluoromethane
5. Chloroethane
6. Methylene chloride
7. Trichlorofluoromethane
8. 1,1-Dichloroethylene
9. Bromochloromethane (IS)
10. 1,1-Dichloroethane
11. trans-1,2-Dichloroethylene
12. Chloroform
13. 1,2-Dichloroethane
14. 1,1,1-Trichloroethane
15. Carbon tetrachloride
16. Bromodichloromethane
17. 1,2-Dichloropropane
18. trans-1,3-Dichloropropene
19. Trichloroethylene
20. Benzene
21. cis-1,3-Dichloropropene,
    1,1,2-Trichloroethane,
    Chlorodibromomethane
22. 1-Chloro-2-bromopropane (IS)
23. 2-Chloroethyl vinyl ether
24. Bromoform
25. 1,1,2,2-Tetrachloroethylene,
    1,1,2,2-Tetrachloroethane
26. 1,4-Dichlorobutane (IS)
27. Toluene
28. Chlorobenzene
29. Ethylbenzene
30. 1,3-Dichlorobenzene
31. 1,2-Dichlorobenzene
32. 1,4-Dichlorobenzene

713-0704

US EPA Method 625, Pesticides

Packing: 3% SP-2250 on 100/120 SUPELCOPORT
Cat. No.: 11878 (packing)
Column: 2m x 2mm ID glass
Oven: 50°C (4 min) to 270°C at 8°C/min
Carrier: nitrogen, 30mL/min
Det.: FID
Inj.: 0.1μL of a synthetic mixture of pesticides (1ng/μL each)

1. α-BHC
2. γ-BHC
3. β-BHC
4. Heptachlor
5. δ-BHC
6. Aldrin
7. Heptachlorepoxide
8. Endosulfan I
9. Dieldrin & 4,4'-DDE
10. Endrin
11. Endosulfan II & 4,4'-DDD
12. 4,4'-DDT
13. Endosulfan sulfate & Endrin aldehyde

795-0193

251

Ethylene Oxide and Ethylene Oxide Residues

Packing: **80/100 Carbopack C/0.8% THEED**
Cat. No.: **11880-U** (packing)
Column: 1 m x 2mm ID glass
(stock column available)
Oven: 115°C
Carrier: nitrogen, 20mL/min
Det.: FID
Inj.: 1μL water containing
50ppm each component

1. Ethylene oxide
2. Water
3. Ethylene chlorohydrin
4. Ethylene glycol

713-0732

Trace Ethylene Oxide in Nitrogen

Packing: **HayeSep D**
Cat. No.: **10292** (packing)
Column: 10 x 1/8" stainless steel
Oven: 130°C
Carrier: helium, 30mL/min
Det.: FID, 140°C
Inj.: 250mL Valco valve, 100°C

1. Ethylene oxide, 23ppm

795-0195

Ethylene Oxide by US Army Method

Sample: HBr-coated Carboxen-564 spiked
with 2-bromoethanol (1.2ng) and
desorbed with toluene:acetonitrile
(50:50)

Sampling Tube: **ORBO-78**
Cat. No.: **20355**
Packing: **10% DEGS on 80/100**
SUPELCOPORT

Cat. No.: **11999** (packing, 20g/bottle)
Column: 12 x 1/8" stainless steel
Oven: 155°C
Carrier: nitrogen, 30mL/min
Det.: ECD

S Solvent
1 Solvent impurities

2-Bromoethanol

EtO Analyses
2-Bromoethanol back-
ground is negligible
Blank

713-0696, 0697

Fatty Acid Methyl Esters of Rapeseed Oil Mixture

Packing: **5% DEGS-PS on 100/120 SUPELCOPORT**
Cat. No.: **11870-U** (packing, 20g/bottle)
Column: 6' x 1/8" stainless steel
Oven: 200°C
Carrier: nitrogen, 20mL/min
Det.: FID
Inj.: 0.5μL RM-3 Mix (Cat. No. O7256-1AMP), 10μg FAMEs/μL in chloroform

1. C14:0
2. C16:0
3. C18:0
4. C18:1
5. C18:2
6. C20:0
7. C18:3
8. C22:0
9. C22:1
10. C24:0

713-0973

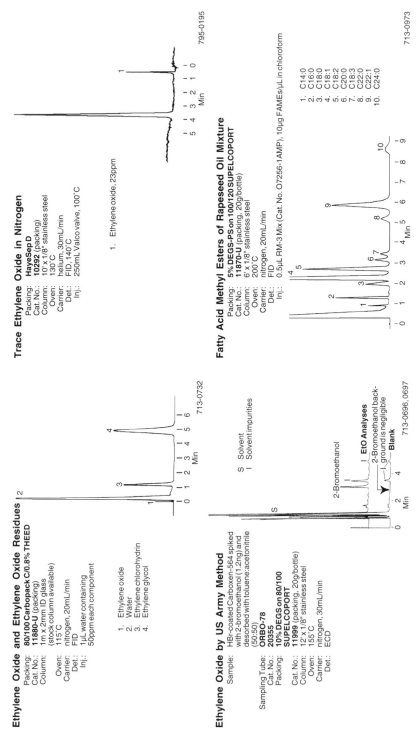

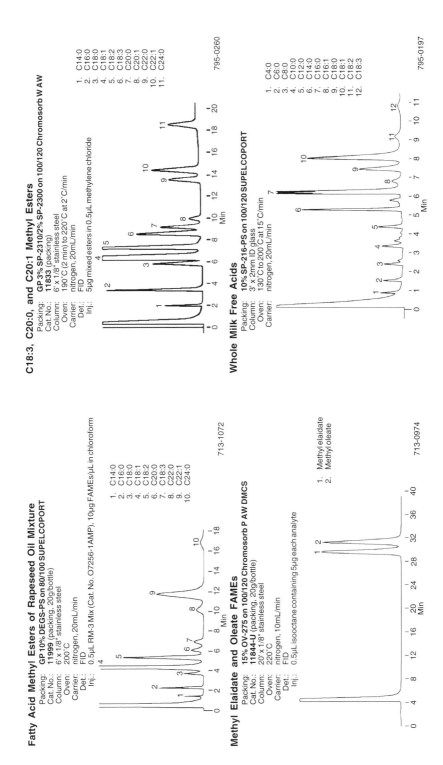

Fatty Acid Methyl Esters of Rapeseed Oil Mixture

Packing: **GP 10% DEGS-PS on 80/100 SUPELCOPORT**
Cat. No.: **11999** (packing, 20g/bottle)
Column: 6' x 1/8" stainless steel
Oven: 200°C
Carrier: nitrogen, 20mL/min
Det.: FID
Inj.: 0.5µL RM-3 Mix (Cat. No. O7256-1 AMP), 10µg FAMEs/µL in chloroform

1. C14:0
2. C16:0
3. C18:0
4. C18:1
5. C18:2
6. C20:0
7. C18:3
8. C22:0
9. C22:1
10. C24:0

713-1072

Methyl Elaidate and Oleate FAMEs

Packing: **15% OV-275 on 100/120 Chromosorb P AW DMCS**
Cat. No.: **11844-U** (packing, 20g/bottle)
Column: 20' x 1/8" stainless steel
Oven: 220°C
Carrier: nitrogen, 10mL/min
Det.: FID
Inj.: 0.5µL isooctane containing 5µg each analyte

1. Methyl elaidate
2. Methyl oleate

713-0974

C18:3, C20:0, and C20:1 Methyl Esters

Packing: **GP 3% SP-2310/2% SP-2300 on 100/120 Chromosorb W AW**
Cat. No.: **11833** (packing)
Column: 6' x 1/8" stainless steel
Oven: 190°C (2 min) to 220°C at 2°C/min
Carrier: nitrogen, 20mL/min
Det.: FID
Inj.: 5µg mixed esters in 0.5µL methylene chloride

1. C14:0
2. C16:0
3. C18:0
4. C18:1
5. C18:2
6. C18:3
7. C20:0
8. C20:1
9. C22:0
10. C22:1
11. C24:0

795-0260

Whole Milk Free Acids

Packing: **10% SP-216-PS on 100/120 SUPELCOPORT**
Column: 3' x 2mm ID glass
Oven: 130°C to 200°C at 15°C/min
Carrier: nitrogen, 20mL/min

1. C4:0
2. C6:0
3. C8:0
4. C10:0
5. C12:0
6. C14:0
7. C16:0
8. C16:1
9. C18:0
10. C18:1
11. C18:2
12. C18:3

795-0197

253

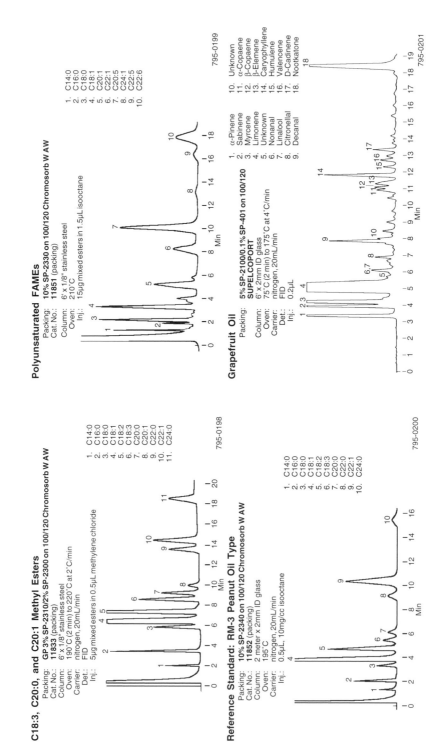

C18:3, C20:0, and C20:1 Methyl Esters

Packing: GP 3% SP-2310/2% SP-2300 on 100/120 Chromosorb W AW
Cat. No.: 11833 (packing)
Column: 6' x 1/8" stainless steel
Oven: 190°C (2 min) to 220°C at 2°C/min
Carrier: nitrogen, 20mL/min
Det.: FID
Inj.: 5µg mixed esters in 0.5µL methylene chloride

1. C14:0
2. C16:0
3. C18:0
4. C18:1
5. C18:2
6. C18:3
7. C20:0
8. C20:1
9. C22:0
10. C22:1
11. C24:0

795-0198

Polyunsaturated FAMEs

Packing: 10% SP-2330 on 100/120 Chromosorb W AW
Cat. No.: 11851 (packing)
Column: 6' x 1/8" stainless steel
Oven: 210°C
Inj.: 15µg mixed esters in 1.5µL isooctane

1. C14:0
2. C16:0
3. C18:0
4. C18:1
5. C20:1
6. C22:1
7. C20:5
8. C24:1
9. C22:5
10. C22:6

795-0199

Reference Standard: RM-3 Peanut Oil Type

Packing: 10% SP-2340 on 100/120 Chromosorb W AW
Cat. No.: 11852 (packing)
Column: 2 meter x 2mm ID glass
Oven: 195°C
Carrier: nitrogen, 20mL/min
Inj.: 0.5µL, 10mg/cc isooctane

1. C14:0
2. C16:0
3. C18:0
4. C18:1
5. C18:2
6. C18:3
7. C20:0
8. C22:0
9. C22:1
10. C24:0

795-0200

Grapefruit Oil

Packing: 5% SP-2100/0.1% SP-401 on 100/120 SUPELCOPORT
Column: 6' x 2mm ID glass
Oven: 75°C (2 min) to 175°C at 4°C/min
Carrier: nitrogen, 20mL/min
Det.: FID
Inj.: 0.2µL

1. α-Pinene
2. Sabinene
3. Myrcene
4. Limonene
5. Unknown
6. Nonanal
7. Linalool
8. Citronellal
9. Decanal
10. Unknown
11. α-Copaene
12. β-Copaene
13. β-Elemene
14. Caryophyllene
15. Humulene
16. Valencene
17. D-Cadinene
18. Nootkatone

795-0201

254

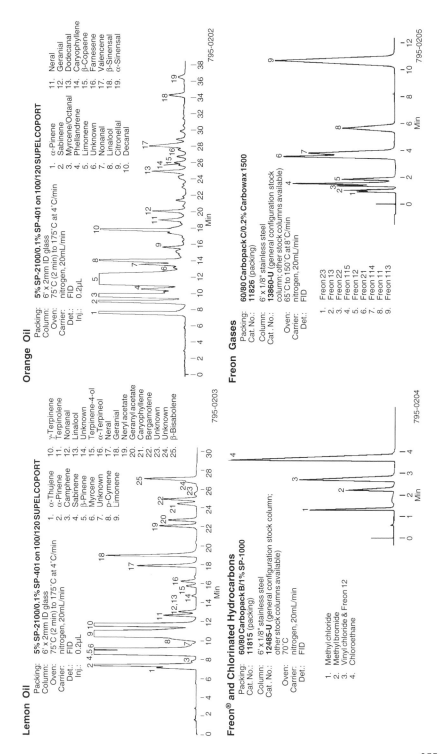

Lemon Oil

Packing: 5% SP-2100/0.1% SP-401 on 100/120 SUPELCOPORT
Column: 6' x 2mm ID glass
Oven: 75°C (2 min) to 175°C at 4°C/min
Carrier: nitrogen, 20mL/min
Det.: FID
Inj.: 0.2µL

1. α-Thujene
2. α-Pinene
3. Camphene
4. Sabinene
5. β-Pinene
6. Myrcene
7. Unknown
8. p-Cymene
9. Limonene
10. γ-Terpinene
11. Terpinolene
12. Nonanal
13. Linalool
14. Unknown
15. Terpinene-4-ol
16. α-Terpineol
17. Neral
18. Geranial
19. Nerylacetate
20. Geranylacetate
21. Caryophyllene
22. Bergamotene
23. Unknown
24. Unknown
25. β-Bisabolene

795-0203

Orange Oil

Packing: 5% SP-2100/0.1% SP-401 on 100/120 SUPELCOPORT
Column: 6' x 2mm ID glass
Oven: 75°C (2 min) to 175°C at 4°C/min
Carrier: nitrogen, 20mL/min
Det.: FID
Inj.: 0.2µL

1. α-Pinene
2. Sabinene
3. Myrcene/Octanal
4. Phellandrene
5. Limonene
6. Unknown
7. Nonanal
8. Linalool
9. Citronellal
10. Decanal
11. Neral
12. Geranial
13. Dodecanal
14. Caryophyllene
15. β-Copaene
16. Farnesene
17. Valencene
18. β-Sinensal
19. α-Sinensal

795-0202

Freon® and Chlorinated Hydrocarbons

Packing: 60/80 Carbopack B/1% SP-1000
Cat. No.: 11815 (packing)
Column: 6' x 1/8" stainless steel
Cat. No.: 12485-U (general configuration stock column; other stock columns available)
Oven: 70°C
Carrier: nitrogen, 20mL/min
Det.: FID

1. Methyl chloride
2. Methyl bromide
3. Vinyl chloride & Freon 12
4. Chloroethane

795-0204

Freon Gases

Packing: 60/80 Carbopack C/0.2% Carbowax 1500
Cat. No.: 11826 (packing)
Column: 6' x 1/8" stainless steel
Cat. No.: 13860-U (general configuration stock column; other stock columns available)
Oven: 65°C to 150°C at 8°C/min
Carrier: nitrogen, 20mL/min
Det.: FID

1. Freon 23
2. Freon 13
3. Freon 22
4. Freon 115
5. Freon 12
6. Freon 21
7. Freon 114
8. Freon 11
9. Freon 113

795-0205

255

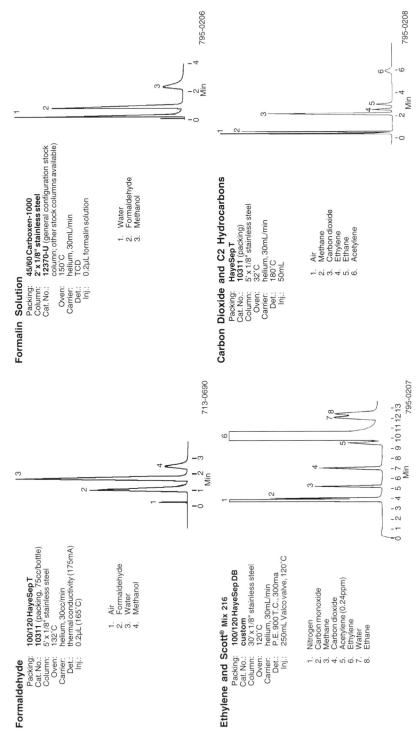

Formaldehyde

Packing: **100/120 HayeSep T**
Cat. No.: **10311** (packing, 75cc/bottle)
Column: 5' x 1/8" stainless steel
Oven: 132°C
Carrier: helium, 30cc/min
Det.: thermal conductivity (175mA)
Inj.: 0.2µL (165°C)

1. Air
2. Formaldehyde
3. Water
4. Methanol

713-0690

Formalin Solution

Packing: **45/60 Carboxen-1000**
Column: **2' x 1/8" stainless steel**
Cat. No.: **12370-U** (general configuration stock column; other stock columns available)
Oven: 150°C
Carrier: helium, 30mL/min
Det.: TCD
Inj.: 0.2µL formalin solution

1. Water
2. Formaldehyde
3. Methanol

795-0206

Ethylene and Scott® Mix 216

Packing: **100/120 HayeSep DB**
custom
Cat. No.: 30 x 1/8" stainless steel
Column: 120°C
Oven: helium, 30mL/min
Carrier: P.E. 900 T.C., 300ma
Det.: 250mL Valco valve, 120°C
Inj.:

1. Nitrogen
2. Carbon monoxide
3. Methane
4. Carbon dioxide
5. Acetylene (0.24ppm)
6. Ethylene
7. Water
8. Ethane

795-0207

Carbon Dioxide and C2 Hydrocarbons

Packing: **HayeSep T**
10311 (packing)
Cat. No.: 5' x 1/8" stainless steel
Column: 32°C
Oven: helium, 30mL/min
Carrier: 180°C
Det.: 50mL
Inj.:

1. Air
2. Methane
3. Carbon dioxide
4. Ethylene
5. Ethane
6. Acetylene

795-0208

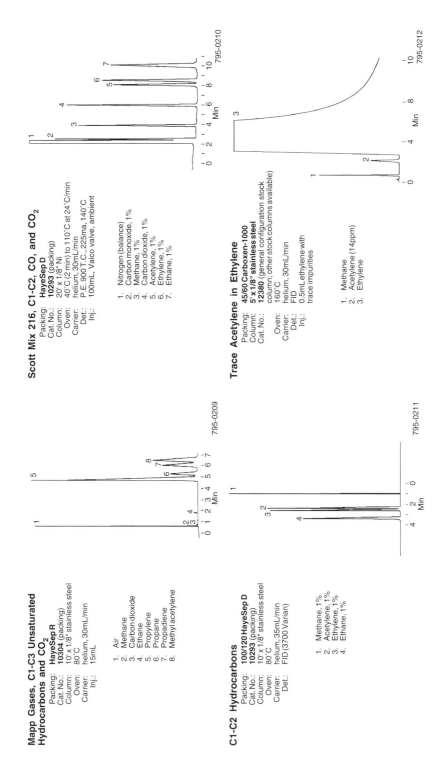

Mapp Gases, C1-C3 Unsaturated Hydrocarbons and CO₂

Packing: **HayeSep R**
Cat. No.: **10304** (packing)
Column: 10' x 1/8" stainless steel
Oven: 80°C
Carrier: helium, 30mL/min
Inj.: 15mL

1. Air
2. Methane
3. Carbon dioxide
4. Ethane
5. Propylene
6. Propane
7. Propadiene
8. Methyl acetylene

795-0209

Scott Mix 216, C1-C2, CO, and CO₂

Packing: **HayeSep D**
Cat. No.: **10293** (packing)
Column: 20' x 1/8" Ni
Oven: 40°C (2 min) to 110°C at 24°C/min
Carrier: helium, 30mL/min
Det.: P.E. 900 T.C., 225ma, 140°C
Inj.: 100mL, Valco valve, ambient

1. Nitrogen (balance)
2. Carbon monoxide, 1%
3. Methane, 1%
4. Carbon dioxide, 1%
5. Acetylene, 1%
6. Ethylene, 1%
7. Ethane, 1%

795-0210

C1-C2 Hydrocarbons

Packing: **100/120 HayeSep D**
Cat. No.: **10293** (packing)
Column: 10' x 1/8" stainless steel
Oven: 80°C
Carrier: helium, 35mL/min
Det.: FID (3700 Varian)

1. Methane, 1%
2. Acetylene, 1%
3. Ethylene, 1%
4. Ethane, 1%

795-0211

Trace Acetylene in Ethylene

Packing: **45/60 Carboxen-1000**
Column: **5' x 1/8" stainless steel**
Cat. No.: **12380** (general configuration stock column; other stock columns available)
Oven: 160°C
Carrier: helium, 30mL/min
Det.: FID
Inj.: 0.5mL ethylene with trace impurities

1. Methane
2. Acetylene (14ppm)
3. Ethylene

795-0212

257

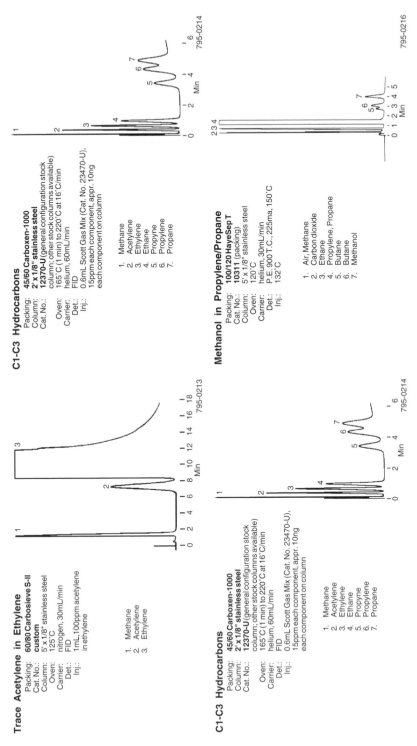

Trace Acetylene in Ethylene

Packing: **60/80 Carbosieve S-II**
custom
Cat. No.:
Column: 5' x 1/8" stainless steel
Oven: 125°C
Carrier: nitrogen, 30mL/min
Det.: FID
Inj.: 1mL, 100ppm acetylene in ethylene

1. Methane
2. Acetylene
3. Ethylene

795-0213

C1-C3 Hydrocarbons

Packing: **45/60 Carboxen-1000**
Column: **2' x 1/8" stainless steel**
Cat. No.: **12370-U** (general configuration stock column; other stock columns available)
Oven: 165°C (1 min) to 220°C at 16°C/min
Carrier: helium, 60mL/min
Det.: FID
Inj.: 0.6mL Scott Gas Mix (Cat. No. 23470-U), 15ppm each component, appr. 10ng each component on column

1. Methane
2. Acetylene
3. Ethylene
4. Ethane
5. Propyne
6. Propylene
7. Propane

795-0214

C1-C3 Hydrocarbons

Packing: **45/60 Carboxen-1000**
Column: **2' x 1/8" stainless steel**
Cat. No.: **12370-U** (general configuration stock column; other stock columns available)
Oven: 165°C (1 min) to 220°C at 16°C/min
Carrier: helium, 60mL/min
Det.: FID
Inj.: 0.6mL Scott Gas Mix (Cat. No. 23470-U), 15ppm each component, appr. 10ng each component on column

1. Methane
2. Acetylene
3. Ethylene
4. Ethane
5. Propyne
6. Propylene
7. Propane

795-0214

Methanol in Propylene/Propane

Packing: **100/120 HayeSep T**
Cat. No.: **10311** (packing)
Column: 5' x 1/8" stainless steel
Oven: 120°C
Carrier: helium, 30mL/min
P.E. 900 T.C., 225ma, 150°C
Det.: 132°C
Inj.:

1. Air, Methane
2. Carbon dioxide
3. Ethane
4. Propylene, Propane
5. Butane
6. Butane
7. Methanol

795-0216

258

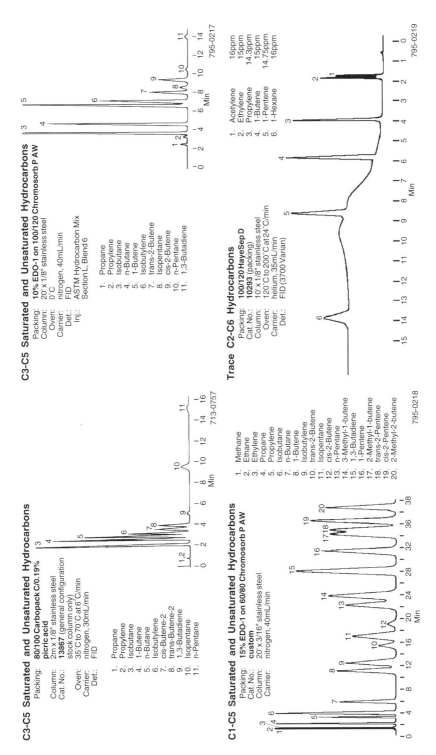

C3-C5 Saturated and Unsaturated Hydrocarbons

Packing: **80/100 Carbopack C/0.19% picric acid**
Column: 2m x 1/8" stainless steel
Cat. No.: **13867** (general configuration stock column only)
Oven: 35°C to 70°C at 6°C/min
Carrier: nitrogen, 30mL/min
Det.: FID

1. Propane
2. Propylene
3. Isobutane
4. 1-Butene
5. n-Butane
6. Isobutylene
7. cis-Butene-2
8. trans-Butene-2
9. 1,3-Butadiene
10. Isopentane
11. n-Pentane

713-0757

C3-C5 Saturated and Unsaturated Hydrocarbons

Packing: **10% EDO-1 on 100/120 Chromosorb P AW**
Column: 20 x 1/8" stainless steel
Oven: 0°C
Carrier: nitrogen, 40mL/min
Det.: FID
Inj.: ASTM Hydrocarbon Mix Section L, Blend 6

1. Propane
2. Propylene
3. Isobutane
4. n-Butane
5. 1-Butene
6. Isobutylene
7. trans-2-Butene
8. Isopentane
9. cis-2-Butene
10. n-Pentane
11. 1,3-Butadiene

795-0217

C1-C5 Saturated and Unsaturated Hydrocarbons

Packing: **15% EDO-1 on 60/80 Chromosorb P AW custom**
Cat. No.:
Column: 20 x 3/16" stainless steel
Carrier: nitrogen, 40mL/min

1. Methane
2. Ethane
3. Ethylene
4. Propane
5. Propylene
6. Isobutane
7. n-Butane
8. 1-Butene
9. Isobutylene
10. trans-2-Butene
11. Isopentane
12. cis-2-Butene
13. n-Pentane
14. 3-Methyl-1-butene
15. 1,3-Butadiene
16. 1-Pentene
17. 2-Methyl-1-butene
18. trans-2-Pentene
19. cis-2-Pentene
20. 2-Methyl-2-butene

795-0218

Trace C2-C6 Hydrocarbons

Packing: **100/120 HayeSep D**
Cat. No.: **10293** (packing)
Column: 10 x 1/8" stainless steel
Oven: 120°C to 200°C at 24°C/min
Carrier: helium, 35mL/min
Det.: FID (3700 Varian)

1. Acetylene   16ppm
2. Ethylene   15ppm
3. Propylene   15ppm
4. 1-Butene   14.3ppm
5. 1-Pentene   14.75ppm
6. 1-Hexane   16ppm

795-0219

259

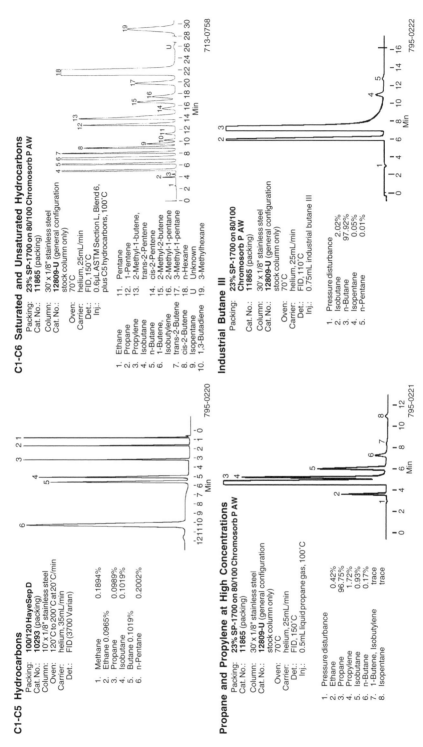

C1-C5 Hydrocarbons

Packing: **100/120 HayeSep D**
Cat. No.: **10293** (packing)
Column: 10' x 1/8" stainless steel
Oven: 120°C to 200°C at 20°C/min
Carrier: helium, 35mL/min
Det.: FID (3700 Varian)

1. Methane 0.1894%
2. Ethane 0.0965%
3. Propane 0.0989%
4. Isobutane 0.1019%
5. Butane 0.1019%
6. n-Pentane 0.2002%

795-0220

C1-C6 Saturated and Unsaturated Hydrocarbons

Packing: **23% SP-1700 on 80/100 Chromosorb P AW**
Cat. No.: **11865** (packing)
Column: 30' x 1/8" stainless steel
Cat. No.: **12809-U** (general configuration stock column only)
Oven: 70°C
Carrier: helium, 25mL/min
Det.: FID, 150°C
Inj.: 0.6μL ASTM Section L, Blend 6, plus C5 hydrocarbons, 100°C

1. Ethane
2. Propane
3. Propylene
4. Isobutane
5. 1-Butene,
6. Isobutylene
7. trans-2-Butene
8. cis-2-Butene
9. Isopentane
10. 1,3-Butadiene

11. Pentane
12. 1-Pentene
13. 2-Methyl-1-butene, trans-2-Pentene
14. cis-2-Pentene
15. 2-Methyl-2-butene
16. 2-Methyl-1-pentene
17. 3-Methyl-1-pentane
18. n-Hexane
U. Unknown
19. 3-Methylhexane

713-0758

Propane and Propylene at High Concentrations

Packing: **23% SP-1700 on 80/100 Chromosorb P AW**
Cat. No.: **11865** (packing)
Column: 30' x 1/8" stainless steel
Cat. No.: **12809-U** (general configuration stock column only)
Oven: 70°C
Carrier: helium, 25mL/min
Det.: FID, 150°C
Inj.: 0.5mL liquid propane gas, 100°C

1. Pressure disturbance
2. Ethane 0.42%
3. Propane 96.75%
4. Propylene 1.72%
5. Isobutane 0.93%
6. n-Butane 0.17%
7. 1-Butene, Isobutylene trace
8. Isopentane trace

795-0221

Industrial Butane III

Packing: **23% SP-1700 on 80/100 Chromosorb P AW**
Cat. No.: **11865** (packing)
Column: 30' x 1/8" stainless steel
Cat. No.: **12809-U** (general configuration stock column only)
Oven: 70°C
Carrier: helium, 25mL/min
Det.: FID, 110°C
Inj.: 0.75mL industrial butane III

1. Pressure disturbance
2. Isobutane 2.02%
3. n-Butane 97.92%
4. Isopentane 0.05%
5. n-Pentane 0.01%

795-0222

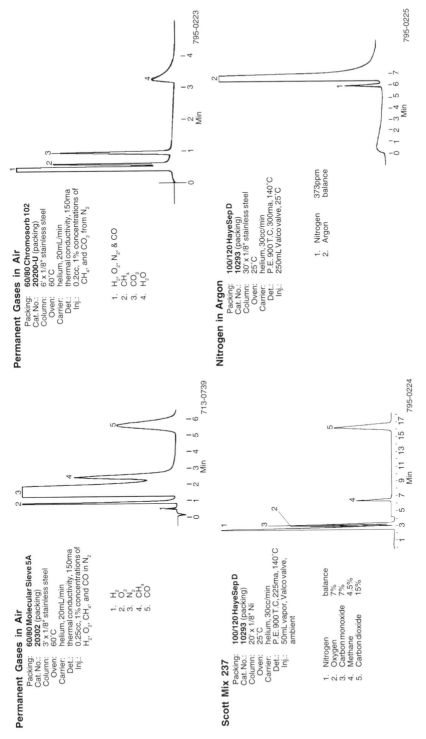

Permanent Gases in Air

Packing: **60/80 Molecular Sieve 5A**
Cat. No.: **20302** (packing)
Column: 3' x 1/8" stainless steel
Oven: 60°C
Carrier: helium, 20mL/min
Det.: thermal conductivity, 150ma
Inj.: 0.25cc, 1% concentrations of H_2, O_2, CH_4, and CO in N_2

1. H_2
2. O_2
3. N_2
4. CH_4
5. CO

713-0739

Permanent Gases in Air

Packing: **60/80 Chromosorb 102**
Cat. No.: **20200-U** (packing)
Column: 6' x 1/8" stainless steel
Oven: 60°C
Carrier: helium, 20mL/min
Det.: thermal conductivity, 150ma
Inj.: 0.2cc, 1% concentrations of CH_4, and CO_2 from N_2

1. H_2, O_2, N_2, & CO
2. CH_4
3. CO_2
4. H_2O

795-0223

Scott Mix 237

Packing: **100/120 HayeSep D**
Cat. No.: **10293** (packing)
Column: 20' x 1/8" Ni
Oven: 25°C
Carrier: helium, 30cc/min
Det.: P.E. 900 T.C. 225ma, 140°C
Inj.: 50mL vapor, Valco valve, ambient

1. Nitrogen   balance
2. Oxygen   7%
3. Carbon monoxide   7%
4. Methane   4.5%
5. Carbon dioxide   15%

795-0224

Nitrogen in Argon

Packing: **100/120 HayeSep D**
Cat. No.: **10293** (packing)
Column: 30' x 1/8" stainless steel
Oven: 25°C
Carrier: helium, 30cc/min
Det.: P.E. 900 T.C. 300ma, 140°C
Inj.: 250mL Valco valve, 25°C

1. Nitrogen   373ppm
2. Argon   balance

795-0225

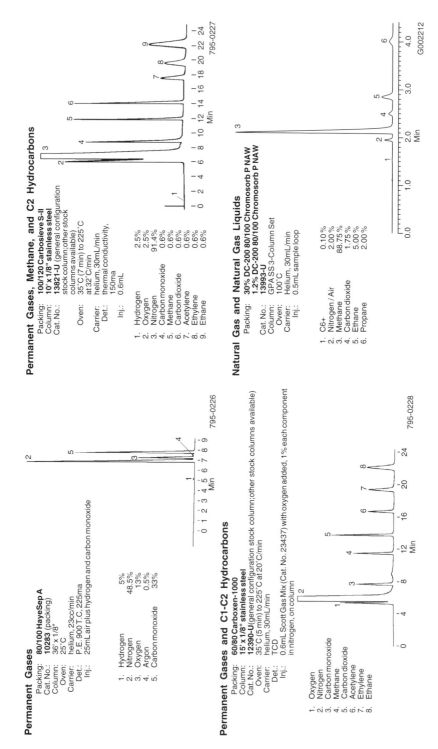

Permanent Gases

Packing: **80/100 HayeSep A**
Cat. No.: **10283 (packing)**
Column: 36' x 1/8"
Oven: 25°C
Carrier: helium, 23cc/min
Det.: P.E. 900 T.C. 225ma
Inj.: 25mL air plus hydrogen and carbon monoxide

1. Hydrogen — 5%
2. Nitrogen — 48.5%
3. Oxygen — 13%
4. Argon — 0.5%
5. Carbon monoxide — 33%

795-0226

Permanent Gases, Methane, and C2 Hydrocarbons

Packing: **100/120 Carbosieve S-II**
Column: **10' x 1/8" stainless steel**
Cat. No.: **13821-U** (general configuration stock column; other stock columns available)
Oven: 35°C (7 min) to 225°C at 32°C/min
Carrier: helium, 30mL/min
Det.: thermal conductivity, 150ma
Inj.: 0.6mL

1. Hydrogen — 2.5%
2. Oxygen — 2.5%
3. Nitrogen — 91.4%
4. Carbon monoxide — 0.6%
5. Methane — 0.6%
6. Carbon dioxide — 0.6%
7. Acetylene — 0.6%
8. Ethylene — 0.6%
9. Ethane — 0.6%

795-0227

Permanent Gases and C1-C2 Hydrocarbons

Packing: **60/80 Carboxen-1000**
Column: **15' x 1/8" stainless steel**
Cat. No.: **12390-U** (general configuration stock column; other stock columns available)
Oven: 35°C (5 min) to 225°C at 20°C/min
Carrier: helium, 30mL/min
Det.: TCD
Inj.: 0.6mL Scott Gas Mix (Cat. No. 23437) with oxygen added, 1% each component in nitrogen, on column

1. Oxygen
2. Nitrogen
3. Carbon monoxide
4. Methane
5. Carbon dioxide
6. Acetylene
7. Ethylene
8. Ethane

795-0228

Natural Gas and Natural Gas Liquids

Packing: **30% DC-200 80/100 Chromosorb P NAW**
1.2% DC-200 80/100 Chromosorb P NAW
Cat. No.: **13993-U**
Column: GPA SS 3-Column Set
Oven: 100°C
Carrier: Helium, 30mL/min
Inj.: 0.5mL sample loop

1. C6+ — 0.10%
2. Nitrogen / Air — 2.00%
3. Methane — 88.75%
4. Carbon dioxide — 1.75%
5. Ethane — 5.00%
6. Propane — 2.00%

G002212

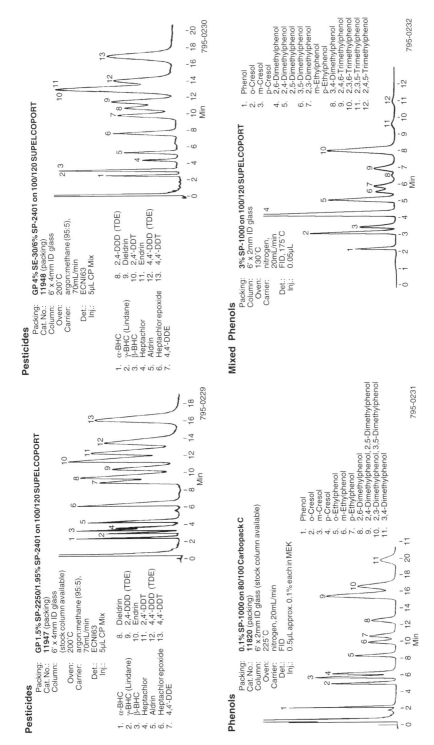

Pesticides

Packing: GP 1.5% SP-2250/1.95% SP-2401 on 100/120 SUPELCOPORT
Cat. No.: 11947 (packing)
Column: 6' x 4mm ID glass
(stock column available)
Oven: 200°C
Carrier: argon:methane (95:5),
70mL/min
Det.: ECNi63
Inj.: 5µL CP Mix

1. α-BHC   8. Dieldrin
2. γ-BHC (Lindane)   9. 2,4-DDD (TDE)
3. β-BHC   10. 2,4'-DDT
4. Heptachlor   11. 2,4'-DDT
5. Aldrin   12. 4,4'-DDD (TDE)
6. Heptachlor epoxide   13. 4,4'-DDT
7. 4,4'-DDE

795-0229

Pesticides

Packing: GP 4% SE-30/6% SP-2401 on 100/120 SUPELCOPORT
Cat. No.: 11948 (packing)
Column: 6' x 4mm ID glass
Oven: 200°C
Carrier: argon:methane (95:5),
70mL/min
Det.: ECNi63
Inj.: 5µL CP Mix

1. α-BHC   8. 2,4-DDD (TDE)
2. γ-BHC (Lindane)   9. Dieldrin
3. β-BHC   10. 2,4'-DDT
4. Heptachlor   11. Endrin
5. Aldrin   12. 4,4'-DDD (TDE)
6. Heptachlor epoxide   13. 4,4'-DDT
7. 4,4'-DDE

795-0230

Phenols

Packing: 0.1% SP-1000 on 80/100 Carbopack C
Cat. No.: 11820 (packing)
Column: 6' x 2mm ID glass (stock column available)
Oven: 225°C
Carrier: nitrogen, 20mL/min
Det.: FID
Inj.: 0.5µL approx. 0.1% each in MEK

1. Phenol
2. o-Cresol
3. m-Cresol
4. p-Cresol
5. o-Ethylphenol
6. m-Ethylphenol
7. p-Ethylphenol
8. 2,6-Dimethylphenol
9. 2,4-Dimethylphenol, 2,5-Dimethylphenol
10. 2,3-Dimethylphenol, 3,5-Dimethylphenol
11. 3,4-Dimethylphenol

795-0231

Mixed Phenols

Packing: 3% SP-1000 on 100/120 SUPELCOPORT
Column: 6' x 2mm ID glass
Oven: 130°C
Carrier: nitrogen,
20mL/min
Det.: FID, 175°C
Inj.: 0.05µL

1. Phenol
2. o-Cresol
3. m-Cresol
4. p-Cresol
5. 2,6-Dimethylphenol
   2,4-Dimethylphenol
   2,5-Dimethylphenol
6. 3,5-Dimethylphenol
7. 2,3-Dimethylphenol
   m-Ethylphenol
   p-Ethylphenol
8. 3,4-Dimethylphenol
9. 2,4,6-Trimethylphenol
10. 2,3,6-Trimethylphenol
11. 2,3,5-Trimethylphenol
12. 2,4,5-Trimethylphenol

795-0232

263

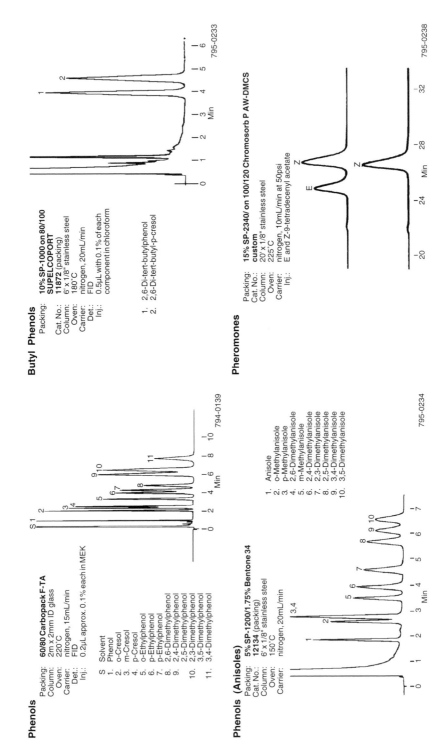

Phenols

Packing: **60/80 Carbopack F-TA**
Column: 2 m x 2 mm ID glass
Oven: 220°C
Carrier: nitrogen, 15 mL/min
Det.: FID
Inj.: 0.2 µL approx. 0.1% each in MEK

S Solvent
1. Phenol
2. o-Cresol
3. m-Cresol
4. p-Cresol
5. o-Ethylphenol
6. p-Ethylphenol
7. p-Ethylphenol
8. 2,6-Dimethylphenol
9. 2,4-Dimethylphenol
    2,5-Dimethylphenol
10. 2,3-Dimethylphenol
    3,5-Dimethylphenol
11. 3,4-Dimethylphenol

794-0139

Butyl Phenols

Packing: **10% SP-1000 on 80/100 SUPELCOPORT**
Cat. No.: **11872** (packing)
Column: 6' x 1/8" stainless steel
Oven: 180°C
Carrier: nitrogen, 20 mL/min
Det.: FID
Inj.: 0.5 µL with 0.1% of each component in chloroform

1. 2,6-Di-tert-butylphenol
2. 2,6-Di-tert-butyl-p-cresol

795-0233

Phenols (Anisoles)

Packing: **5% SP-1200/1.75% Bentone 34**
Cat. No.: **12134** (packing)
Column: 6' x 1/8" stainless steel
Oven: 150°C
Carrier: nitrogen, 20 mL/min

1. Anisole
2. o-Methylanisole
3. p-Methylanisole
4. 2,6-Dimethylanisole
5. m-Methylanisole
6. 2,4-Dimethylanisole
7. 2,3-Dimethylanisole
8. 2,5-Dimethylanisole
9. 3,4-Dimethylanisole
10. 3,5-Dimethylanisole

795-0234

Pheromones

Packing: **15% SP-2340/ on 100/120 Chromosorb P AW-DMCS custom**
Cat. No.:
Column: 20' x 1/8" stainless steel
Oven: 225°C
Carrier: nitrogen, 10 mL/min at 50 psi
Inj.: E and Z-9-tetradecenyl acetate

795-0238

264

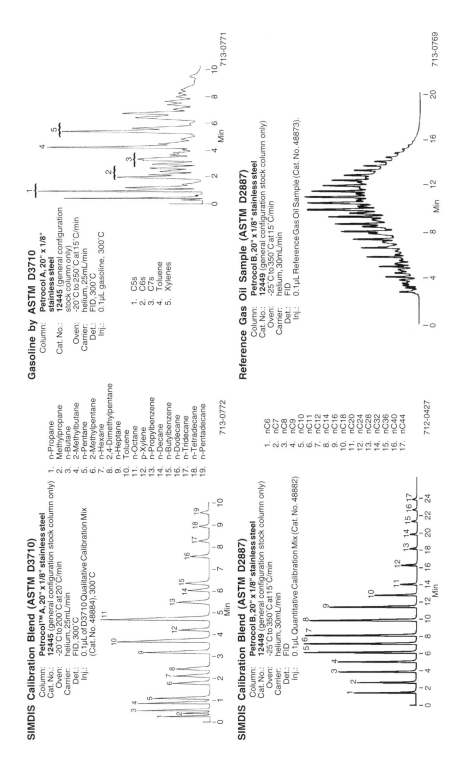

SIMDIS Calibration Blend (ASTM D3710)

Column: **Petrocol™ A, 20" x 1/8" stainless steel**
Cat. No.: **12445** (general configuration stock column only)
Oven: -20°C to 200°C at 20°C/min
Carrier: helium, 25mL/min
Det.: FID, 300°C
Inj.: 0.1µL of D3710 Qualitative Calibration Mix
(Cat. No. 48884), 300°C

713-0772

1. n-Propane
2. Methylpropane
3. n-Butane
4. 2-Methylbutane
5. n-Pentane
6. 2-Methylpentane
7. n-Hexane
8. 2,4-Dimethylpentane
9. n-Heptane
10. Toluene
11. n-Octane
12. p-Xylene
13. n-Propylbenzene
14. n-Decane
15. n-Butylbenzene
16. n-Dodecane
17. n-Tridecane
18. n-Tetradecane
19. n-Pentadecane

SIMDIS Calibration Blend (ASTM D2887)

Column: **Petrocol B, 20" x 1/8" stainless steel**
Cat. No.: **12449** (general configuration stock column only)
Oven: -25°C to 350°C at 15°C/min
Carrier: helium, 30mL/min
Det.: FID
Inj.: 0.1µL Quantitative Calibration Mix (Cat. No. 48882)

712-0427

1. nC6
2. nC7
3. nC8
4. nC9
5. nC10
6. nC11
7. nC12
8. nC14
9. nC16
10. nC18
11. nC20
12. nC24
13. nC28
14. nC32
15. nC36
16. nC40
17. nC44

Gasoline by ASTM D3710

Column: **Petrocol A, 20" x 1/8" stainless steel**
Cat. No.: **12445** (general configuration stock column only)
Oven: -20°C to 250°C at 15°C/min
Carrier: helium, 25mL/min
Det.: FID, 300°C
Inj.: 0.1µL gasoline, 300°C

1. C5s
2. C6s
3. C7s
4. Toluene
5. Xylenes

713-0771

Reference Gas Oil Sample (ASTM D2887)

Column: **Petrocol B, 20" x 1/8" stainless steel**
Cat. No.: **12449** (general configuration stock column only)
Oven: -25°C to 350°C at 15°C/min
Carrier: helium, 30mL/min
Det.: FID
Inj.: 0.1µL Reference Gas Oil Sample (Cat. No. 48873).

713-0769

265

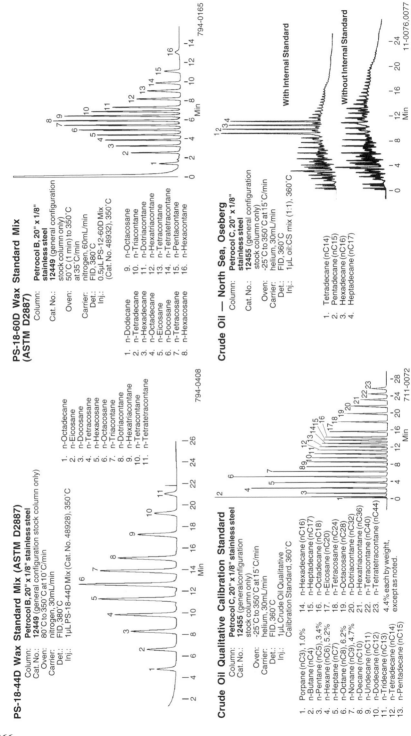

PS-18-44D Wax Standard Mix (ASTM D2887)

Column: **Petrocol B, 20" x 1/8" stainless steel**
Cat. No.: **12449** (general configuration stock column only)
Oven: 80°C to 350°C at 10°C/min
Carrier: nitrogen, 30mL/min
Det.: FID, 380°C
Inj.: 1µL PS-18-44D Mix (Cat. No. 48928), 350°C

1. n-Octadecane
2. n-Eicosane
3. n-Docosane
4. n-Tetracosane
5. n-Hexacosane
6. n-Octacosane
7. n-Triacontane
8. n-Dotriacontane
9. n-Hexatriacontane
10. n-Tetracontane
11. n-Tetratetracontane

794-0408

PS-18-60D Wax Standard Mix (ASTM D2887)

Column: **Petrocol B, 20" x 1/8" stainless steel**
Cat. No.: **12449** (general configuration stock column only)
Oven: 50°C (1 min) to 350°C at 35°C/min
Carrier: nitrogen, 60mL/min
Det.: FID, 380°C
Inj.: 0.5µL PS-12-60D Mix (Cat. No. 48932), 350°C

1. n-Dodecane
2. n-Tetradecane
3. n-Hexadecane
4. n-Octadecane
5. n-Eicosane
6. n-Docosane
7. n-Tetracosane
8. n-Hexacosane
9. n-Octacosane
10. n-Triacontane
11. n-Dotriacontane
12. n-Hexatriacontane
13. n-Tetracontane
14. n-Tetratetracontane
15. n-Pentacontane
16. n-Hexacontane

794-0165

Crude Oil Qualitative Calibration Standard

Column: **Petrocol C, 20" x 1/8" stainless steel**
Cat. No.: **12455** (general configuration stock column only)
Oven: -25°C to 350°C at 15°C/min
Carrier: helium, 30mL/min
Det.: FID, 360°C
Inj.: 1µL Crude Oil Qualitative Calibration Standard, 360°C

1. Porpane (nC3), 1.0%
2. n-Butane (nC4)
3. n-Pentane (nC5), 3.4%
4. n-Hexane (nC6), 5.2%
5. n-Heptane (nC7)
6. n-Octane (nC8), 6.2%
7. n-Nonane (nC9), 4.7%
8. n-Decane (nC10)
9. n-Undecane (nC11)
10. n-Dodecane (nC12)
11. n-Tridecane (nC13)
12. n-Tetradecane (nC14)
13. n-Pentadecane (nC15)
14. n-Hexadecane (nC16)
15. n-Heptadecane (nC17)
16. n-Octadecane (nC18)
17. n-Eicosane (nC20)
18. n-Tetracosane (nC24)
19. n-Octacosane (nC28)
20. n-Dotriacontane (nC32)
21. n-Hexatriacontane (nC36)
22. n-Tetracontane (nC40)
23. n-Tetratetracontane (nC44)

4.4% each by weight, except as noted.

711-0072

Crude Oil — North Sea, Oseberg

Column: **Petrocol C, 20" x 1/8" stainless steel**
Cat. No.: **12455** (general configuration stock column only)
Oven: -25°C to 350°C at 15°C/min
Carrier: helium, 30mL/min
Det.: FID, 360°C
Inj.: 1µL oil:CS mix (1:1), 360°C

1. Tetradecane (nC14)
2. Pentadecane (nC15)
3. Hexadecane (nC16)
4. Heptadecane (nC17)

With Internal Standard

Without Internal Standard

11-0076,0077

266

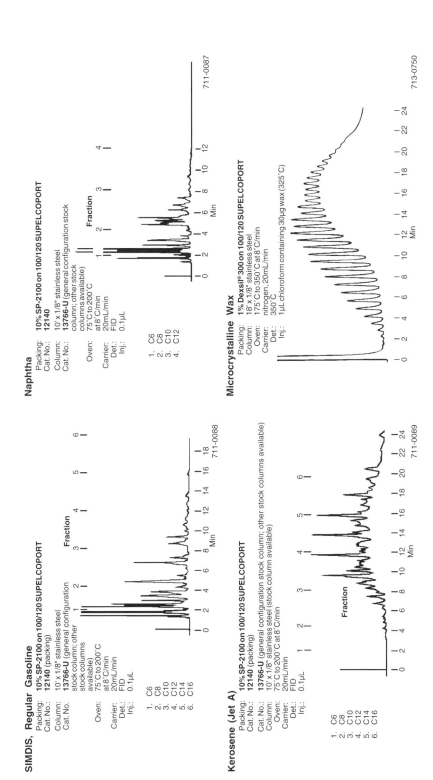

SIMDIS, Regular Gasoline

Packing: **10% SP-2100 on 100/120 SUPELCOPORT**
Cat. No.: **12140** (packing)
Column: 10' x 1/8" stainless steel
Cat. No. **13766-U** (general configuration stock columns; other stock columns available)
Oven: 75°C to 200°C at 8°C/min
Carrier: 20mL/min
Det.: FID
Inj.: 0.1μL

1. C6
2. C8
3. C10
4. C12
5. C14
6. C16

711-0088

Naphtha

Packing: **10% SP-2100 on 100/120 SUPELCOPORT**
Cat. No.: **12140**
Column: 10' x 1/8" stainless steel
Cat. No.: **13766-U** (general configuration stock column; other stock columns available)
Oven: 75°C to 200°C at 8°C/min
Carrier: 20mL/min
Det.: FID
Inj.: 0.1μL

1. C6
2. C8
3. C10
4. C12

711-0087

Kerosene (Jet A)

Packing: **10% SP-2100 on 100/120 SUPELCOPORT**
Cat. No.: **12140** (packing)
Cat. No.: **13766-U** (general configuration stock column; other stock columns available)
Column: 10' x 1/8" stainless steel (stock column available)
Oven: 75°C to 200°C at 8°C/min
Carrier: 20mL/min
Det.: FID
Inj.: 0.1μL

1. C6
2. C8
3. C10
4. C12
5. C14
6. C16

711-0089

Microcrystalline Wax

Packing: **1% Dexsil® 300 on 100/120 SUPELCOPORT**
Column: 18' x 1/8" stainless steel
Oven: 175°C to 350°C at 8°C/min
Carrier: nitrogen, 20mL/min
Det.: 350°C
Inj.: 1μL chloroform containing 30μg wax (325°C)

713-0750

267

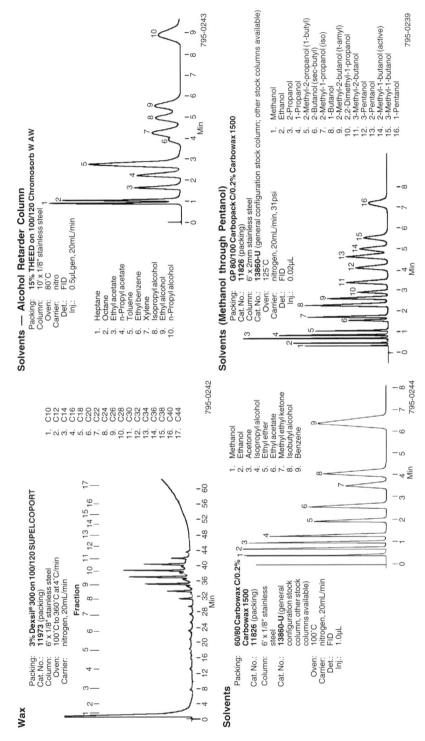

268

Wax

Packing: **3% Dexsil® 300 on 100/120 SUPELCOPORT**
Cat. No.: **11973 (packing)**
Column: 6' x 1/8" stainless steel
Oven: 100°C to 360°C at 4°C/min
Carrier: nitrogen, 20mL/min

Fraction

1. C10
2. C12
3. C14
4. C16
5. C18
6. C20
7. C22
8. C24
9. C26
10. C28
11. C30
12. C32
13. C34
14. C36
15. C38
16. C40
17. C44

795-0242

Solvents — Alcohol Retarder Column

Packing: **15% THEED on 100/120 Chromosorb W AW**
Column: 10' x 1/8" stainless steel
Oven: 80°C
Carrier: nitro
Det.: FID
Inj.: 0.5µL gen, 20mL/min

1. Heptane
2. Octane
3. Ethyl acetate
4. n-Propyl acetate
5. Toluene
6. Ethyl benzene
7. Xylene
8. Isopropyl alcohol
9. Ethyl alcohol
10. n-Propyl alcohol

795-0243

Solvents

Packing: **60/80 Carbowax C/0.2%**
Carbowax 1500
Cat. No.: **11826 (packing)**
Column: 6' x 1/8" stainless steel
Cat. No.: **13860-U** (general configuration stock column, other stock columns available)
Oven: 100°C
Carrier: nitrogen, 20mL/min
Det.: FID
Inj.: 1.0µL

1. Methanol
2. Ethanol
3. Acetone
4. Isopropyl alcohol
5. Ethyl ether
6. Ethyl acetate
7. Methyl ethyl ketone
8. Isobutyl alcohol
9. Benzene

795-0244

Solvents (Methanol through Pentanol)

Packing: **GP 80/100 Carbopack C/0.2% Carbowax 1500**
Cat. No.: **11826 (packing)**
Column: 6' x 2mm stainless steel
Cat. No.: **13860-U** (general configuration stock column; other stock columns available)
Oven: 125°C
Carrier: nitrogen, 20mL/min, 31psi
Det.: FID
Inj.: 0.02µL

1. Methanol
2. Ethanol
3. 2-Propanol
4. 1-Propanol
5. 2-Methyl-2-propanol (1-butyl)
6. 2-Butanol (sec-butyl)
7. 2-Methyl-1-propanol (iso)
8. 1-Butanol
9. 2-Methyl-2-butanol (t-amyl)
10. 2,2-Dimethyl-1-propanol
11. 3-Methyl-2-butanol
12. 3-Pentanol
13. 2-Pentanol
14. 2-Methyl-1-butanol (active)
15. 3-Methyl-1-butanol
16. 1-Pentanol

795-0239

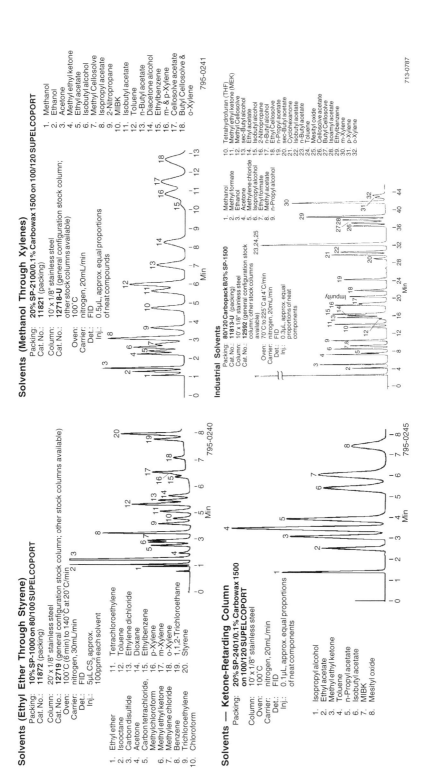

Solvents (Ethyl Ether Through Styrene)

Packing: 10% SP-1000 on 80/100 SUPELCOPORT
Cat. No.: 11872 (packing)
Column: 20' x 1/8" stainless steel
Cat. No.: 12719 (general configuration stock column; other stock columns available)
Oven: 100°C (6 min) to 140°C at 20°C/min
Carrier: nitrogen, 30mL/min
Det.: FID
Inj.: 5µL CS₂, approx.
100ppm each solvent

1. Ethyl ether
2. Isooctane
3. Carbon disulfide
4. Acetone
5. Carbon tetrachloride,
   Methylchloroform
6. Methyl ethyl ketone
7. Methylene chloride
8. Benzene
9. Trichloroethylene
10. Chloroform
11. Tetrachloroethylene
12. Toluene
13. Ethylene dichloride
14. Dioxane
15. Ethylbenzene
   p-Xylene
16. m-Xylene
17. o-Xylene
18. Benzene
19. 1,1,2-Trichloroethane
20. Styrene

795-0240

Solvents — Ketone-Retarding Column

Packing: 20% SP-2401/0.1% Carbowax 1500 on 100/120 SUPELCOPORT
Column: 10' x 1/8" stainless steel
Oven: 100°C
Carrier: nitrogen, 20mL/min
Det.: FID
Inj.: 0.1µL, approx. equal proportions of neat components

1. Isopropyl alcohol
2. Ethyl acetate
3. Methyl ethyl ketone
4. Toluene
5. n-Propyl acetate
6. Isobutyl acetate
7. MIBK
8. Mesityl oxide

795-0245

Solvents (Methanol Through Xylenes)

Packing: 20% SP-2100/0.1% Carbowax 1500 on 100/120 SUPELCOPORT
Cat. No.: 11821 (packing)
Column: 10' x 1/8" stainless steel
Cat. No.: 12718-U (general configuration stock column; other stock columns available)
Oven: 100°C
Carrier: nitrogen, 20mL/min
Det.: FID
Inj.: 0.5µL, approx. equal proportions of neat compounds

1. Methanol
2. Ethanol
3. Acetone
4. Methyl ethyl ketone
5. Ethyl acetate
6. Isobutyl alcohol
7. Methyl Cellosolve
8. Isopropyl acetate
9. 2-Nitropropane
10. MIBK
11. Isobutyl acetate
12. Toluene
13. n-Butyl acetate
14. Diacetone alcohol
15. Ethylbenzene
16. m- & p-Xylene
17. Cellosolve acetate
18. Butyl Cellosolve & o-Xylene

713-0787

Industrial Solvents

Packing: 80/120 Carbopack B/3% SP-1500
Cat. No.: 11813-U (packing)
Column: 10' x 1/8" stainless steel
Cat. No.: 12592 (general configuration stock column; other stock columns available)
Oven: 70°C to 225°C at 4°C/min
Carrier: nitrogen, 20mL/min
Det.: FID
Inj.: 0.3µL, approx. equal proportions of neat components

1. Methanol
2. Methyl formate
3. Ethanol
4. Acetone
5. Methylene chloride
6. Isopropyl alcohol
7. Ethyl formate
8. Methyl acetate
9. n-Propyl alcohol
10. Tetrahydrofuran (THF)
11. Methyl ethyl ketone (MEK)
12. Methyl Cellosolve
13. sec-Butyl alcohol
14. Ethyl acetate
15. Isopropyl alcohol
16. 2-Nitropropane
17. n-Butyl alcohol
18. Ethyl Cellosolve
19. n-Propyl acetate
20. sec-Butyl acetate
21. Isobutyl acetate
22. Cyclohexanone
23. Isobutyl acetate
24. n-Butyl acetate
25. Toluene
26. Mesityl oxide
27. Cellosolve acetate
28. Butyl Cellosolve
29. Isoamyl acetate
30. Ethylbenzene
31. m-Xylene
32. o-Xylene

795-0241

269

Industrial Solvents

Packing: 60/80 Carbopack B/1% SP-1510
Cat.No.: 11809 (packing)
Column: 10' x 1/8" stainless steel
Oven: 100°C to 225°C at 8°C/min
Carrier: nitrogen, 20mL/min
Det.: FID
Inj.: 0.3µL, approx. equal proportions of neat components

1. Methanol
2. Methyl formate
3. Ethanol
4. Methylene chloride
5. Acetone
6. Isopropyl alcohol
7. Ethyl formate
8. Methyl acetate
9. n-Propyl alcohol
10. Tetrahydrofuran (THF)
11. Methyl ethyl ketone (MEK)
12. Methyl Cellosolve
13. sec-Butyl alcohol
14. Isobutyl alcohol
15. Ethyl acetate
16. 2-Nitropropane
17. n-Butyl alcohol
18. Ethyl Cellosolve
19. n-Propyl acetate
20. Cyclohexanone
21. sec-Butyl acetate
22. Isobutyl acetate
23. n-Butyl acetate
24. Mesityl oxide
25. Toluene
26. Cellosolve acetate
27. Butyl Cellosolve
28. Isoamyl acetate
29. Ethylbenzene
30. m-Xylene
31. o-Xylene
32. p-Xylene

713-0788

Industrial Solvents

Packing: 60/80 Carbopack F-TA
Column: 2m x 2mm ID glass
Oven: 75°C (2 min) to 225°C at 4°C/min
Carrier: nitrogen, 20mL/min
Det.: FID
Inj.: 0.05µL solvents mix, 3% each solvent

1. Methanol
2. Methyl formate
3. Ethanol
4. Acetone
5. Methylene chloride
6. Isopropyl alcohol
7. Ethyl formate
8. Methyl acetate
9. n-Propyl alcohol
10. Tetrahydrofuran (THF)
11. Methyl ethyl ketone (MEK)
12. Methyl Cellosolve
13. sec-Butyl alcohol
14. Ethyl acetate
15. 2-Nitropropane
16. n-Butyl alcohol
I. Impurity
18. Ethyl Cellosolve
19. n-Propyl acetate
20. Cyclohexanone
21. sec-Butyl acetate
22. n-Butyl acetate
23. n-Butyl acetate
24. Toluene
25. Mesityl oxide
26. Cellosolve acetate
27. Butyl Cellosolve
28. Isoamyl acetate
29. Ethylbenzene
30. m-Xylene
31. o-Xylene
32. p-Xylene

795-0248

Industrial Solvents

Packing: 80/100 Carbopack C/0.1% SP-1000
Cat.No.: 11820 (packing)
Column: 10' x 1/8" stainless steel
Oven: 70°C to 225°C at 4°C/min
Carrier: nitrogen, 20mL/min
Det.: FID
Inj.: 0.3µL, 3% each component by volume (approx. 12µg each)

1. Methanol
2. Methyl formate
3. Ethanol
4. Acetone
5. Methylene chloride
6. Isopropyl alcohol
7. Ethyl formate
8. Methyl acetate
9. Tetrahydrofuran (THF)
10. n-Propyl alcohol
11. Methyl ethyl ketone (MEK)
12. Methyl Cellosolve
13. sec-Butyl alcohol
14. Isobutyl alcohol
15. Ethyl acetate
16. 2-Nitropropane
17. n-Butyl alcohol
18. Ethyl Cellosolve
19. n-Propyl acetate
20. Cyclohexanone
21. sec-Butyl acetate
22. Isobutyl acetate
23. n-Butyl acetate
24. Toluene
25. Mesityl oxide
26. Cellosolve acetate
27. Butyl Cellosolve
28. Isoamyl acetate
29. Ethylbenzene
30. m-Xylene
31. p-Xylene
32. o-Xylene

713-0789

Alcohols, C1-C5

Packing: 80/100 Carbopack F-SL
Column: 2m x 2mm ID TightSpec glass
Oven: 130°C
Carrier: nitrogen, 15mL/min

1. Methanol
2. Ethanol
3. i-Propanol
4. n-Propanol
5. t-Butanol
6. i-Butanol
7. 2-Methyl-1-propanol
8. n-Butanol
9. 2-Methyl-2-butanol
10. 2,2-Dimethyl-1-propanol
11. 3-Methyl-2-butanol
12. 3-Methyl-2-butanol
13. 2-Pentanol
14. 2-Methyl-1-butanol
15. 3-Methyl-1-butanol
16. n-Pentanol

713-1069

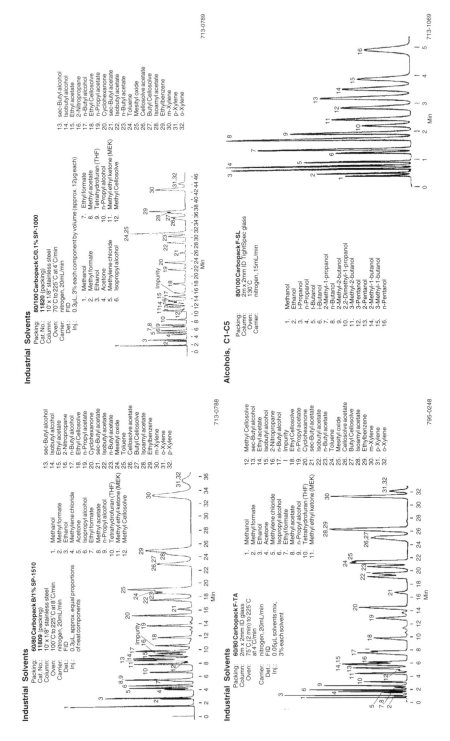

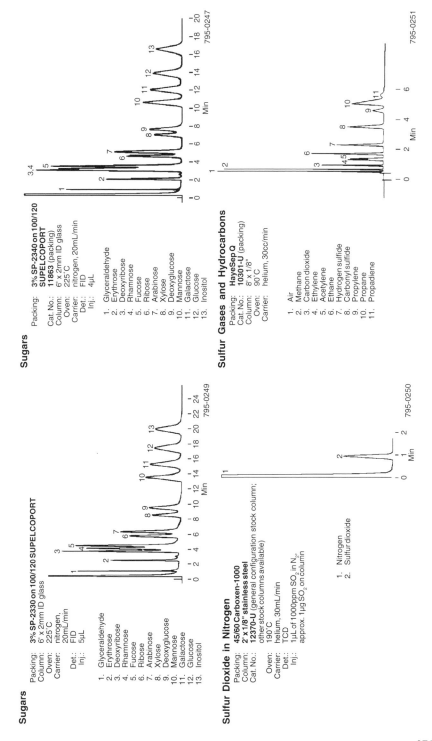

Sugars

Packing: **3% SP-2330 on 100/120 SUPELCOPORT**
Column: 6' x 2mm ID glass
Oven: 225°C
Carrier: nitrogen, 20mL/min
Det.: FID
Inj.: 5μL

1. Glyceraldehyde
2. Erythrose
3. Deoxyribose
4. Rhamnose
5. Fucose
6. Ribose
7. Arabinose
8. Xylose
9. Deoxyglucose
10. Mannose
11. Galactose
12. Glucose
13. Inositol

795-0249

Sugars

Packing: **3% SP-2340 on 100/120 SUPELCOPORT**
Cat. No.: **11863** (packing)
Column: 6' x 2mm ID glass
Oven: 225°C
Carrier: nitrogen, 20mL/min
Det.: FID
Inj.: 4μL

1. Glyceraldehyde
2. Erythrose
3. Deoxyribose
4. Rhamnose
5. Fucose
6. Ribose
7. Arabinose
8. Xylose
9. Deoxyglucose
10. Mannose
11. Galactose
12. Glucose
13. Inositol

795-0247

Sulfur Dioxide in Nitrogen

Packing: **45/60 Carboxen-1000**
Column: **2' x 1/8" stainless steel**
Cat. No.: **12370-U** (general configuration stock column;
other stock columns available)
Oven: 190°C
Carrier: helium, 30mL/min
Det.: TCD
Inj.: 1μL of 1000ppm SO_2 in N_2,
approx. 1μg SO_2 on column

1. Nitrogen
2. Sulfur dioxide

795-0250

Sulfur Gases and Hydrocarbons

Packing: **HayeSep Q**
Cat. No.: **10301-U** (packing)
Column: 8' x 1/8"
Oven: 90°C
Carrier: helium, 30cc/min

1. Air
2. Methane
3. Carbon dioxide
4. Ethylene
5. Acetylene
6. Ethane
7. Hydrogen sulfide
8. Carbonyl sulfide
9. Propylene
10. Propane
11. Propadiene

795-0251

271

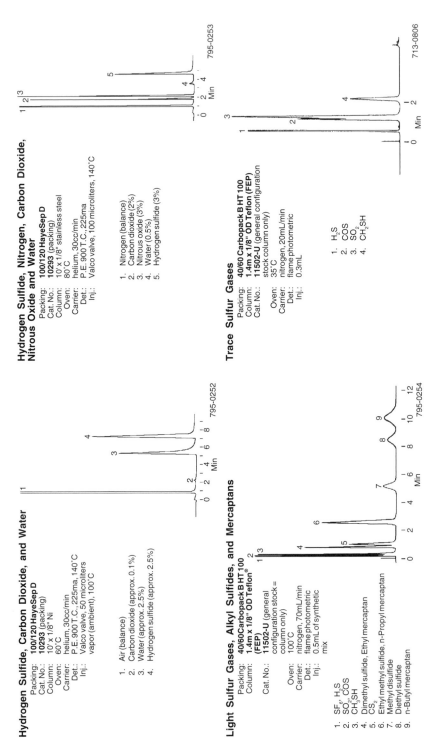

Hydrogen Sulfide, Carbon Dioxide, and Water

Packing: **100/120 HayeSep D**
Cat. No.: **10293** (packing)
Column: 10' x 1/8" Ni
Oven: 60°C
Carrier: helium, 30cc/min
Det.: P.E. 900 T.C., 225ma, 140°C
Inj.: Valco valve, 50 microliters
vapor (ambient), 100°C

1. Air (balance)
2. Carbon dioxide (approx. 0.1%)
3. Water (approx. 2.5%)
4. Hydrogen sulfide (approx. 2.5%)

795-0252

Hydrogen Sulfide, Nitrogen, Carbon Dioxide, Nitrous Oxide and Water

Packing: **100/120 HayeSep D**
Cat. No.: **10293** (packing)
Column: 10' x 1/8" stainless steel
Oven: 80°C
Carrier: helium, 30cc/min
Det.: P.E. 900 T.C., 225ma
Inj.: Valco valve, 100 microliters, 140°C

1. Nitrogen (balance)
2. Carbon dioxide (2%)
3. Nitrous oxide (3%)
4. Water (0.5%)
5. Hydrogen sulfide (3%)

795-0253

Light Sulfur Gases, Alkyl Sulfides, and Mercaptans

Packing: **40/60 Carbopack B HT 100**
Column: **1.4m x 1/8" OD Teflon®**
(FEP)
Cat. No.: **11502-U** (general
configuration stock =
column only)
Oven: 100°C
Carrier: nitrogen, 70mL/min
Det.: flame photometric
Inj.: 0.5mL of synthetic
mix

1. SF₆, H₂S
2. SO₂, COS
3. CH₃SH
4. Dimethyl sulfide, Ethyl mercaptan
5. CS₂
6. Ethyl methyl sulfide, n-Propyl mercaptan
7. Methyl disulfide
8. Diethyl sulfide
9. n-Butyl mercaptan

795-0254

Trace Sulfur Gases

Packing: **40/60 Carbopack B HT 100**
Column: **1.4m x 1/8" OD Teflon (FEP)**
Cat. No.: **11502-U** (general configuration
stock column only)
Oven: 35°C
Carrier: nitrogen, 20mL/min
Det.: flame photometric
Inj.: 0.3mL

1. H₂S
2. COS
3. SO₂
4. CH₂SH

713-0806

272

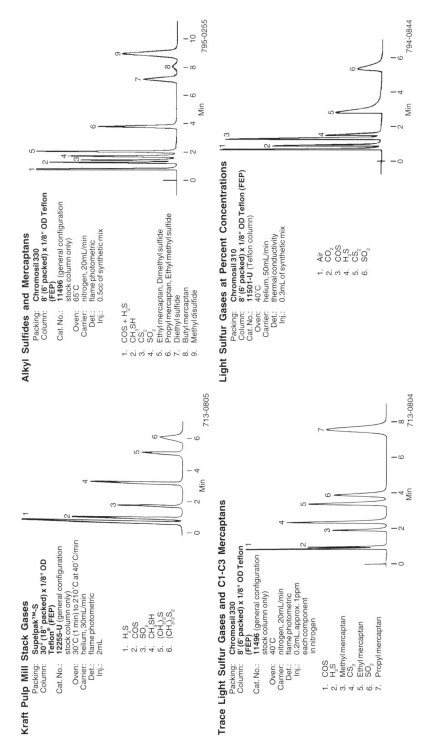

Kraft Pulp Mill Stack Gases

Packing: **Supelpak™-S**
Column: **30" (18" packed) x 1/8" OD Teflon® (FEP)**
Cat. No.: **12255-U** (general configuration stock column only)
Oven: 30°C (1 min) to 210°C at 40°C/min
Carrier: helium, 30mL/min
Det.: flame photometric
Inj.: 2mL

1. H₂S
2. COS
3. SO₂
4. CH₃SH
5. (CH₃)₂S
6. (CH₃)₂S₂

713-0805

Alkyl Sulfides and Mercaptans

Packing: **Chromosil 330**
Column: **8' (6' packed) x 1/8" OD Teflon (FEP)**
Cat. No.: **11496** (general configuration stock column only)
Oven: 65°C
Carrier: nitrogen, 20mL/min
Det.: flame photometric
Inj.: 0.5cc of synthetic mix

1. COS + H₂S
2. CH₃SH
3. CS₂
4. SO₂
5. Ethyl mercaptan, Dimethyl sulfide
6. Propyl mercaptan, Ethyl methyl sulfide
7. Diethyl sulfide
8. Butyl mercaptan
9. Methyl disulfide

795-0255

Trace Light Sulfur Gases and C1-C3 Mercaptans

Packing: **Chromosil 330**
Column: **8' (6' packed) x 1/8" OD Teflon (FEP)**
Cat. No.: **11496** (general configuration stock column only)
Oven: 40°C
Carrier: nitrogen, 20mL/min
Det.: flame photometric
Inj.: 0.2mL, approx. 1ppm each component in nitrogen

1. COS
2. H₂S
3. Methyl mercaptan
4. CS₂
5. Ethyl mercaptan
6. SO₂
7. Propyl mercaptan

713-0804

Light Sulfur Gases at Percent Concentrations

Packing: **Chromosil 310**
Column: **8' (6' packed) x 1/8" OD Teflon (FEP)**
Cat. No.: **11501-U** (Teflon column)
Oven: 40°C
Carrier: helium, 50mL/min
Det.: thermal conductivity
Inj.: 0.3mL of synthetic mix

1. Air
2. CO₂
3. COS
4. H₂S
5. CS₂
6. SO₂

794-0844

273

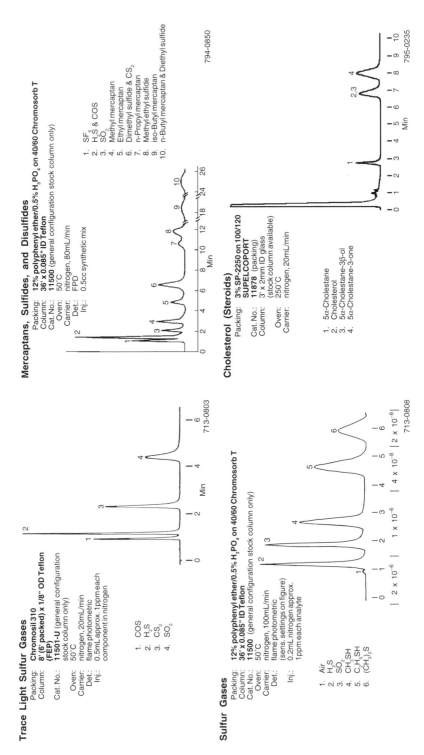

Trace Light Sulfur Gases

Packing: **Chromosil 310**
Column: **8' (6' packed) x 1/8" OD Teflon (FEP)**
Cat. No.: **11501-U** (general configuration stock column only)
Oven: 50°C
Carrier: nitrogen, 20mL/min
Det.: flame photometric
Inj.: 0.5mL approx. 1ppm each component in nitrogen

1. COS
2. H₂S
3. CS₂
4. SO₂

713-0803

Sulfur Gases

Packing: **12% polyphenyl ether/0.5% H₃PO₄ on 40/60 Chromosorb T**
Column: **36' x 0.085" ID Teflon**
Cat. No.: **11500** (general configuration stock column only)
Oven: 50°C
Carrier: nitrogen, 100mL/min
Det.: flame photometric (sens. settings on figure)
Inj.: 0.2mL nitrogen approx. 1ppm each analyte

1. Air
2. H₂S
3. SO₂
4. CH₃SH
5. C₂H₅SH
6. (CH₃)₂S

713-0808

Mercaptans, Sulfides, and Disulfides

Packing: **12% polyphenyl ether/0.5% H₃PO₄ on 40/60 Chromosorb T**
Column: **36' x 0.085" ID Teflon**
Cat. No.: **11500** (general configuration stock column only)
Oven: 50°C
Carrier: nitrogen, 80mL/min
Det.: FPD
Inj.: 0.5cc synthetic mix

1. SF₆
2. H₂S & COS
3. SO₂
4. Methyl mercaptan
5. Ethyl mercaptan
6. Dimethyl sulfide & CS₂
7. n-Propyl mercaptan
8. Methyl ethyl sulfide
9. iso-Butyl mercaptan
10. n-Butyl mercaptan & Diethyl sulfide

794-0850

Cholesterol (Steroids)

Packing: **3% SP-2250 on 100/120 SUPELCOPORT**
Cat. No.: **11878** (packing)
Column: **3' x 2mm ID glass** (stock column available)
Oven: 250°C
Carrier: nitrogen, 20mL/min

1. 5α-Cholestane
2. Cholesterol
3. 5α-Cholestane-3β-ol
4. 5α-Cholestane-3-one

795-0235

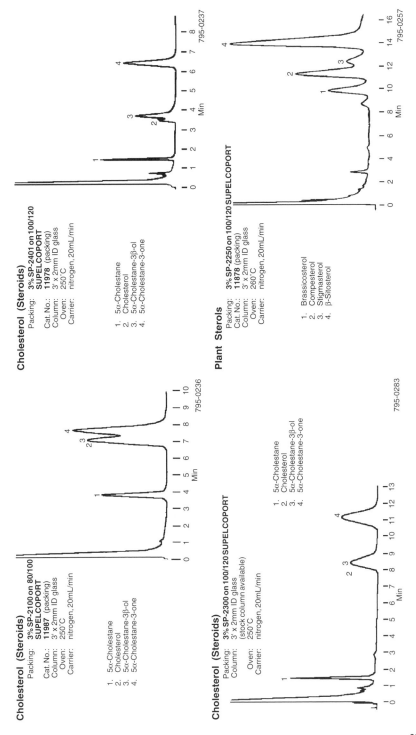

Cholesterol (Steroids)

Packing: **3% SP-2100 on 80/100 SUPELCOPORT**
Cat. No.: **11987** (packing)
Column: 3' x 2mm ID glass
Oven: 250°C
Carrier: nitrogen, 20mL/min

1. 5α-Cholestane
2. Cholesterol
3. 5α-Cholestane-3β-ol
4. 5α-Cholestane-3-one

795-0236

Cholesterol (Steroids)

Packing: **3% SP-2401 on 100/120 SUPELCOPORT**
Cat. No.: **11978** (packing)
Column: 3' x 2mm ID glass
Oven: 250°C
Carrier: nitrogen, 20mL/min

1. 5α-Cholestane
2. Cholesterol
3. 5α-Cholestane-3β-ol
4. 5α-Cholestane-3-one

795-0237

Cholesterol (Steroids)

Packing: **3% SP-2300 on 100/120 SUPELCOPORT**
Column: 3' x 2mm ID glass
(stock column available)
Oven: 250°C
Carrier: nitrogen, 20mL/min

1. 5α-Cholestane
2. Cholesterol
3. 5α-Cholestane-3β-ol
4. 5α-Cholestane-3-one

795-0283

Plant Sterols

Packing: **3% SP-2250 on 100/120 SUPELCOPORT**
Cat. No.: **11878** (packing)
Column: 3' x 2mm ID glass
Oven: 260°C
Carrier: nitrogen, 20mL/min

1. Brassicosterol
2. Compesterol
3. Stigmasterol
4. β-Sitosterol

795-0257

275

Ketosteroids

Packing: **3% OV-225 on 80/100 SUPELCOPORT**
Cat. No.: **11957-U** (packing)
Column: 6' x 4mm ID glass
Oven: 250°C
Carrier: nitrogen, 20mL/min

1. n-Octacosane
2. Androsterone TMS
3. Etiocholanolone TMS
4. Dehydroepiandrosterone TMS
5. 11-Ketoetiocholanolone TMS
6. n-Hexatriacontane
7. 11β-OH-Etiocholanolone TMS

Triglycerides

Packing: **1% Dexsil 300 on 100/120 SUPELCOPORT**
Column: 18" x 1/8" stainless steel
Oven: 275°C to 350°C at 8°C/min
Carrier: nitrogen, 20mL/min
Det.: FID, 350°C
Inj.: 1μL chloroform containing 1μg each triglyceride, 325°C

1. Trilaurin
2. Trimyristin
3. Tripalmitin
4. Tristearin

Pentaerithritol Esters

Packing: **1% Dexsil 300 on 100/120 SUPELCOPORT**
Column: 18" x 1/8" stainless steel
Oven: 125°C to 300°C at 8°C/min
Carrier: nitrogen, 20mL/min
Det.: 350°C

C34 to C48 Alcohols

Cholesteryl Esters

Packing: **1% Dexsil 300 on 100/120 SUPELCOPORT**
Column: 18" x 2mm ID glass
Oven: 300°C to 350°C at 6°C/min
Carrier: nitrogen, 40mL/min
Det.: 350°C

1. Laurate
2. Myristate
3. Palmitate
4. Stearate

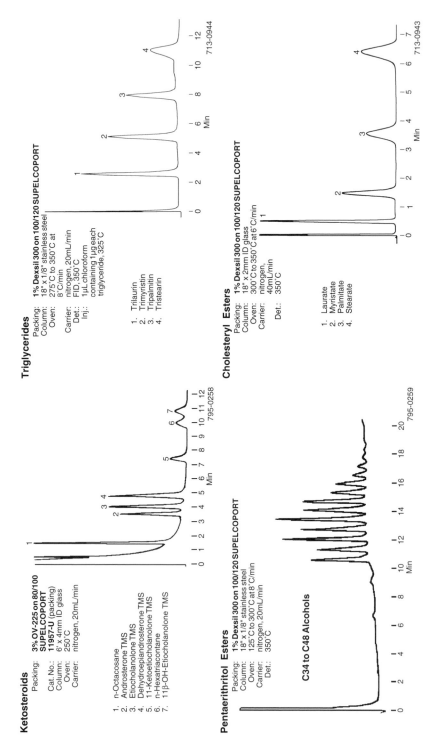

795-0258
713-0944
795-0259
713-0943

APPENDIX B
Column Selection

Selecting the appropriate column for the separation of components in a sample falls into two main categories:

1. All the components of the sample are known and a procedure exists for the separation.
2. All the components of the sample are not known, and thus a procedure may or may not exist for the separation.

In the first category, column selection is straightforward. Search the vendor's catalogs, the journal literature, the chromatograms in Appendix A, or the Internet for separations of samples containing the analyte(s) of interest and follow the gas chromatographic conditions given. In the second category, column selection requires more information and decision making on the part of the chromatographer. A number of questions need to be answered.

> *Important Note:* Chromatographic analysis alone is not a good analytical technique for qualitative analysis.

1. What is the end requirement of the sample analysis?
2. Do you only have to identify a particular analyte or analytes in the sample qualitatively?
3. Do you have to completely separate all components in the sample? *Note:* If the sample is a complete unknown, the number of peaks obtained on the chromatogram does not necessarily furnish the answer. The number of peaks only furnishes information regarding the possible number of components in the sample. You may not have resolved all components in the sample (i.e., some components may have coeluted or fractioned during the separation process).
4. Do you need qualitative and quantitative data for every component of the sample?

Columns for Gas Chromatography: Performance and Selection, by Eugene F. Barry and Robert L. Grob
Copyright © 2007 John Wiley & Sons, Inc.

5. What quantity of sample is available? Are milligram or gram quantities available? If the quantity of sample remaining is critical to the sample submitted, a nondestructive detector must be used.

Important Note: The use of absolute retention time, t_R, to compare an unknown peak to a known peak can be very misleading, as several compounds may have the same absolute retention time on that particular column under those operating conditions of temperature and flow rate.

If the components of the sample are unknown, a good place to start would be with the column you have in your gas chromatograph. If the column is nonpolar, it is more "forgiving." Nonpolar phases are resistant to traces of oxygen and/or water in the carrier gas. Thus, nonpolar phases are more resistant to oxidation and/or hydrolysis than are polar phases. Nonpolar phases are more inert, bleed less, have wider ranges of column operating temperatures, and have higher coating efficiencies. Another property of nonpolar phases is that separations usually occur on the basis of boiling points of sample components.

If not satisfied with the separation, consider known facts about the sample. The underlying principle of chromatographic separations is that sample analytes like to interact with stationary phases similar in chemical properties (i.e., *likes dissolve likes*).

Consider the polar properties of your analyte molecules.

- *Nonpolar.* These are molecules containing only carbon and hydrogen and having no dipole moment. Halocarbons, sulfides, and mercaptans may be included.
- *Weakly intermediate polar.* These may include ethers, aldehydes, ketones, nitro and nitrile compounds (without α-hydrogen atoms), and tertiary amines.
- *Strongly intermediate polar.* These may include alcohols, phenols, primary and secondary amines, carboxylic acids, oximes, and nitro and nitrile compounds (with α-hydrogen atoms).
- *Strongly polar.* These may include amino alcohols, hydroxyl acids, polyhydroxyl alcohols, polyprotic acids, polyphenols, thiols, and molecules containing nitrogen, oxygen, phosphorus, sulfur, or halogen atoms.
- *Polarizable.* These are molecules composed of carbon and hydrogen but also containing unsaturated bonds; (e.g., alkenes, alkynes, aromatic compounds, carbon disulfide).

Important Note: For descriptions of various phases, their uses, and vendors, see Tables 3.13 and 3.14.

Do you require a capillary or a packed gas chromatographic column for your sample analysis? The properties of these two types of columns can be summarized as follows:

Packed Columns	Capillary Columns
High sample capacity	Large-bore capillaries (i.e., 530-μm
Gas sample analysis (however, many recommended packed column procedures are used for samples other than gases)	capillaries have narrowed the advantage of high-capacity packed columns; also, the availability of more sensitive detectors has reduced the requirement for
VOC analysis by headspace sampling technique	large samples)
Purge-and-trap sampling technique	More efficient (i.e., narrower peaks)
	Improved peak separation
	Shortened runtime
	Decreased sample retention
	More samples per unit time

For new or updated methods, the authors recommend capillary columns unless there is an overwhelming reason(s) for using a packed column.

Important Note: As columns age, the stationary phase tends to bake off. This exposes silanol groups (Si–O–H) and peak tailing results. Also, degradation of the stationary phase results from oxygen exposure at high temperatures.

There are other parameters to be considered when selecting a column.

Column Material.   If a capillary column is to be used, fused silica is the material of choice. A number of fused-silica capillary columns are available. The wall-coated open-tubular column (WCOT), where the liquid stationary phase is coated on the deactivated inside wall of the capillary, has become the most widely used type of capillary column. A second type of capillary column is the porous-layer open-tubular column (PLOT), in which the stationary phase consists of solid particles coated on the deactivated inside wall of the capillary. A third type of capillary column is the support-coated open-tubular column (SCOT), in which the liquid stationary phase is coated over the solid particles coated on the deactivated inside wall of the capillary.

If a packed column is to be used, the tubing may be glass or metal (usually stainless steel). Metal columns are more rigid and durable. They are best employed for nonpolar samples. However, if the samples are polar, glass would be a better choice. When using a glass column, peak tailing or sample loss occur may occur; thus, deactivate (silylation treatment) the glass tubing first.

Stationary-Phase Film Thickness.   Film thicknesses between 0.25 and 0.5 μm, the most commonly used, they work well for most sample components eluting up to 300°C. Thinner films (0.1 μm) are recommended for sample components eluting at higher temperatures. Stationary-phase films of 1.0 to 1.5 μm are best for sample components eluting from 100 to 200°C; 3.0 to 5.0 μm is usually required for gases, solvents, and purgeable components. These thick films increase the interaction of sample components and stationary phase. Wide-bore capillary columns are usually available with these thicker films.

Column Inside Diameter.   Larger-diameter columns give more stationary phase, even with the same film thickness; thus greater sample capacity. The negative result means reduced resolving power and greater bleed. Small-diameter columns provide good resolution for complex samples but require a split injection, due to the low sample capacity. Large-i.d. columns or even PLOT columns may be required for the separation of gases, very volatile sample components, and for purge and trap or headspace sampling. It is wise to assess your complete chromatographic system to learn which parameters limit your column diameter choice.

Column Length.   For simple mixtures, fast screening of samples, or high-molecular-weight compounds, the usual column length is 15 m. The popular column length for most analyses is 30 m. Complete separation of complex samples usually requires extra long columns (i.e., 50, 60, or 105 m). Wisdom would dictate that column length is not a consideration for column performance. Doubling the column length doubles isothermal analysis time but only increases peak resolution approximately 40%. If a separation is not quite satisfactory, there are better ways to improve it (e.g., a thinner stationary phase, optimization of the mobile phase flow through the column, or temperature programming). For the analysis of samples containing extremely active components, severe tailing will result if the components contact the column material or uncoated solid support material. Short columns with thick films will reduce interaction by having less solid support particles and column wall material exposed.

INDEX